Breakaway Math

From Practice to Performance

D

Brian Enright, Ed.D.

Joan Fox

Robert Gyles, Ph.D.

Maxine Leonescu

Fred I. Remer

Options Publishing Inc.

P.O. Box 1749
Merrimack, NH 03054-1749
Toll Free: 800-782-7300
Fax: 866-424-4056
www.optionspublishing.com

Table of Contents

Acknowledgments

Executive Editor: Linda Bullock

Editor: Joshua Fisher

Production Supervisor: Sandy Batista

Production Specialist: Corrine Scanlon

Design and Production: Design 5 Creatives

Cover Design: Design 5 Creatives

Illustration Credits: Joe Boddy, Molly K. Scanlon

Photo Credits: Cover and title page: John Kelly/Getty Images, 7: Courtesy of Winchell's Donut House, 11: David Aubrey/Corbis, 13: Chris Hellier/Corbis, 17: SW Productions/Getty Images, 23: Photos.com, 25: John B. Forrest/Corbis, 29: Bettmann/Corbis, 31: Sean Dougherty/Reuters/Corbis, 35: Chris Hellier/Corbis, 41: Large Binocular Telescope Observatory, 47: Roger Garwood/Corbis, 53: National Archives and Records Administration, Records of the Bureau of Public Roads, 59: Photos.com, 65: Will & Deni McIntyre/Getty Images, 71: Mrs. B. W. Riley Pieced quilt, 1939 cotton, pieced 80x75 in. (203.2 x 190.5 cm) Dallas Museum of Art, anonymous gift, 77: Adalberto Rios Szalay/Sexto Sol/Getty Images, 79: National Archives of Australia, 83: Lucy Nicholson/Corbis, 85: Reuters/Corbis, 89: Roger De La Harpe/Corbis, 95: Dan Morphy, 101: Nick Koudis/Getty Images, 107: "Penny images from the United States Mint.", 113: Kelly-Mooney Photography/Corbis, 119: Gary Braasch/Corbis, 125: Lester Lefkowitz/Corbis, 131: Kelly-Mooney Photography/Corbis

ISBN 1-59137-374-3

Printed in the U.S.A.

15 14 13 12 11 10 9 8 7

Where Did the Dough Go?

Who made the doughnut hole? No one knows for sure. One legend says that in 1847, Elizabeth Gregory sent cakes to her son, a sea captain. The captain couldn't hold the cake and steer his ship at the same time. So he pushed the cake over a spoke on the ship's steering wheel. Another story says the captain didn't like nuts and punched them out of the cake's center. Is the doughnut an unusual ocean treasure?

Get Started

Patty and Sean visited the bakery to buy doughnuts for a class party. Draw a picture of the doughnuts on each tray. Write a multiplication fact to describe how many are on each tray. The first one has been done for you.

3 rows of doughnuts
with 5 in each row.

$$\underline{\quad 3 \quad} \times \underline{\quad 5 \quad} = \underline{\quad 15 \quad}$$

2 rows of doughnuts
with 5 in each row.

$$\underline{\qquad} \times \underline{\qquad} = \underline{\qquad}$$

4 rows of doughnuts
with 6 in each row.

$$\underline{\qquad} \times \underline{\qquad} = \underline{\qquad}$$

Working with Partial Products

You can use what you know about multiplication to multiply
numbers with more than one digit.

The example below shows how to multiply a 2-digit number
by a 1-digit number. First multiply the ones. Then multiply
the tens. Each result is called a **partial product.**

$$
\begin{array}{r}
1\ 2 \\
\times\ \ \ 3 \\
\hline
\end{array}
$$

Multiply the ones: $3 \times 2 = 6$.　　6 ← partial products

Multiply the tens: $3 \times 10 = 30$.　$+\ 30$

To find the answer, add the partial products:
$6 + 30 = 36$, so $3 \times 12 = 36$.

The example below shows how to multiply a 3-digit number
by a 1-digit number. Multiply the ones, then the tens, then
the hundreds. Then add the partial products.

$$
\begin{array}{r}
2\ 3\ 5 \\
\times\ \ \ \ \ 4 \\
\hline
\end{array}
$$

Multiply the ones: $4 \times 5 = 20$.　　　　20 ← partial products

Multiply the tens: $4 \times 30 = 120$.　　120

Multiply the hundreds: $4 \times 200 = 800$.　$+800$

Add the partial products. ⟶

1. What is 4×235? _____________

For each multiplication problem below, find the partial products. Then add the partial products to find the answer. Show your work.

2.
$$\begin{array}{r} 56 \\ \times\ 3 \\ \hline \end{array}$$

3.
$$\begin{array}{r} 142 \\ \times\ \ 6 \\ \hline \end{array}$$

4.
$$\begin{array}{r} 28 \\ \times\ 4 \\ \hline \end{array}$$

5.
$$\begin{array}{r} 243 \\ \times\ \ 2 \\ \hline \end{array}$$

Solve a Problem

6. Fill in the boxes at the right to tell what multiplication problem is represented by the model below.

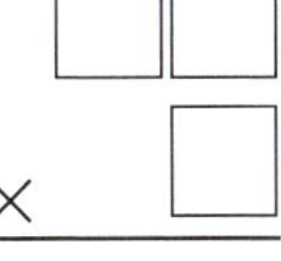

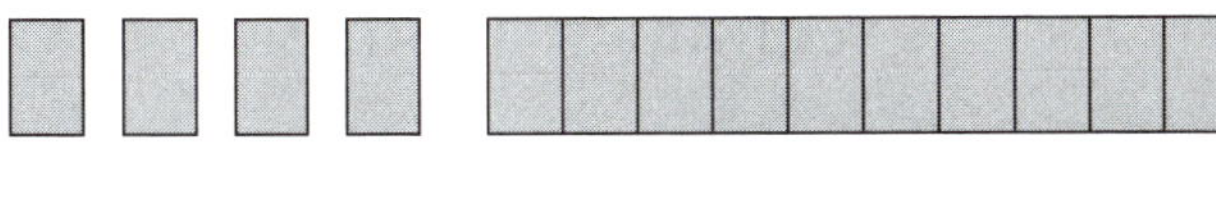

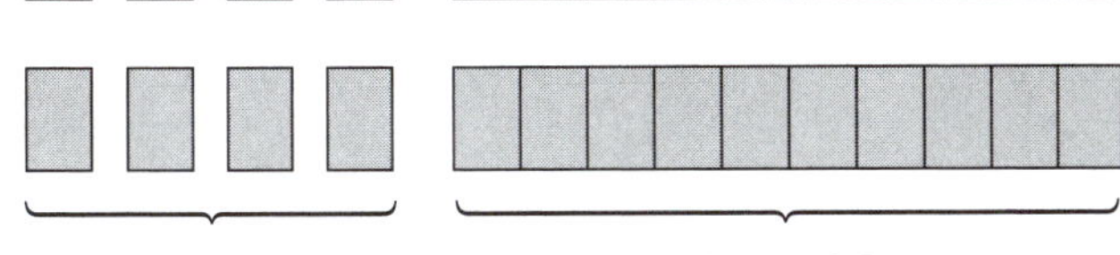

$$2 \times 4 \quad + \quad 2 \times 10$$

It's a Fact!

The largest doughnut ever made weighed 5,000 pounds and stood 95 feet tall.

Multiply with Regrouping

During their bake sale, Byron's class sold 123 homemade doughnuts each day for 5 days. How many doughnuts did they sell in all?

To solve this problem, you can use regrouping to find 5×123. Look at the steps below.

STEP 1 Multiply the ones: 5×3 ones = 15 ones.
Regroup if you need to.

Hundreds	Tens	Ones
	1	
1	2	3
×		5
		5

Regroup 15 ones as 1 ten and 5 ones.

STEP 2 Multiply the tens: 5×2 tens = 10 tens.
Then add the regrouped tens: 10 tens + 1 ten = 11 tens.
Regroup if you need to.

Regroup 11 tens as 1 hundred and 1 ten.

Hundreds	Tens	Ones
1	1	
1	2	3
×		5
	1	5

STEP 3 Multiply the hundreds and add the regrouped hundreds.

Hundreds	Tens	Ones
1	1	
1	2	3
×		5
6	1	5

So, $5 \times 123 =$ __________.

Follow the steps you learned. Use regrouping to solve the multiplication problems below. Show your work.

1.

Hundreds	Tens	Ones
1	2	5
		3

× (Ones, 3)

2.

Hundreds	Tens	Ones
1	0	4
		5

× (Ones, 5)

3.

Hundreds	Tens	Ones
1	5	3
		3

× (Ones, 3)

4.

Hundreds	Tens	Ones
1	6	3
		6

× (Ones, 6)

5.

Hundreds	Tens	Ones
2	0	0
		4

× (Ones, 4)

6.

Hundreds	Tens	Ones
1	9	5
		5

× (Ones, 5)

7. There are 24 hours in one day. Multiply to find the number of hours in one week. Show your work below.

On Your Own

Find the product.

$$159 \times 9$$

Show your work.

There are _________ hours in one week.

1. Multiply.

$$\begin{array}{r} 38 \\ \times\ 1 \\ \hline \end{array}$$

Ⓐ 122

Ⓑ 123

Ⓒ 143

Ⓓ 152

2. What is 3×135?

Ⓕ 405

Ⓖ 315

Ⓗ 300

Ⓙ 138

3. What is the value of 6×70?

Ⓐ 42

Ⓑ 402

Ⓒ 420

Ⓓ 4,200

4. The class library has 8 shelves with 24 books on each shelf. How many books are there in all? Show your work.

____

5. Solve this multiplication problem by adding partial products.

$$\begin{array}{r} 73 \\ \times\ 6 \\ \hline \end{array}$$

6. Think Back How many inches are there in 9 feet? Explain how you found your answer.

9 feet = _______ inches

____

____

____

A Deadly Plant

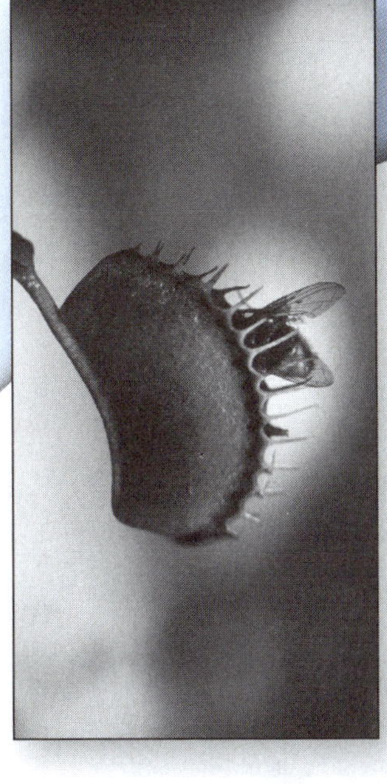

An insect flies toward a cluster of white flowers. Leaves beneath the flowers open like a book. Each open leaf has three tiny hairs. The insect touches a hair, and suddenly, the leaves snap shut. The insect is trapped. The plant begins to ooze a liquid that breaks down the insect's body. After its meal, the Venus flytrap opens its leaves again. It doesn't need to move to catch another insect dinner.

Get Started

Members of the Garden Club planted flower seeds. For each garden, draw one way the gardeners could plant the seeds so that each row has the same number of plants. Then write a division sentence that describes the garden. The first one has been done for you as an example.

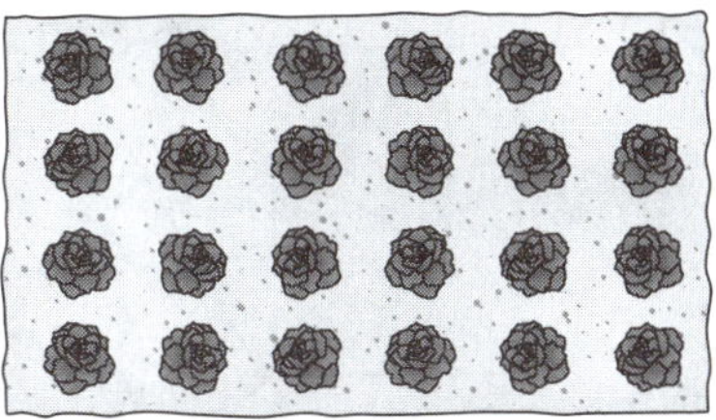

Marlene planted 24 seeds.

$$\underline{\ \ 24\ \ } \div \underline{\ \ 4\ \ } = \underline{\ \ 6\ \ }$$

dividend   divisor   quotient

Ryan planted 16 seeds.

_______ ÷ _______ = _______

Marisol planted 36 seeds.

_______ ÷ _______ = _______

Working with Division

You can use drawings to help you divide with numbers that have more than one digit.

Suppose you want to divide 73 by 2. This means you want to divide 7 tens and 3 ones into 2 equal groups.

First, divide the 7 tens into 2 equal groups.

You can put 3 tens in each group.

You have 1 ten and 3 ones left over.

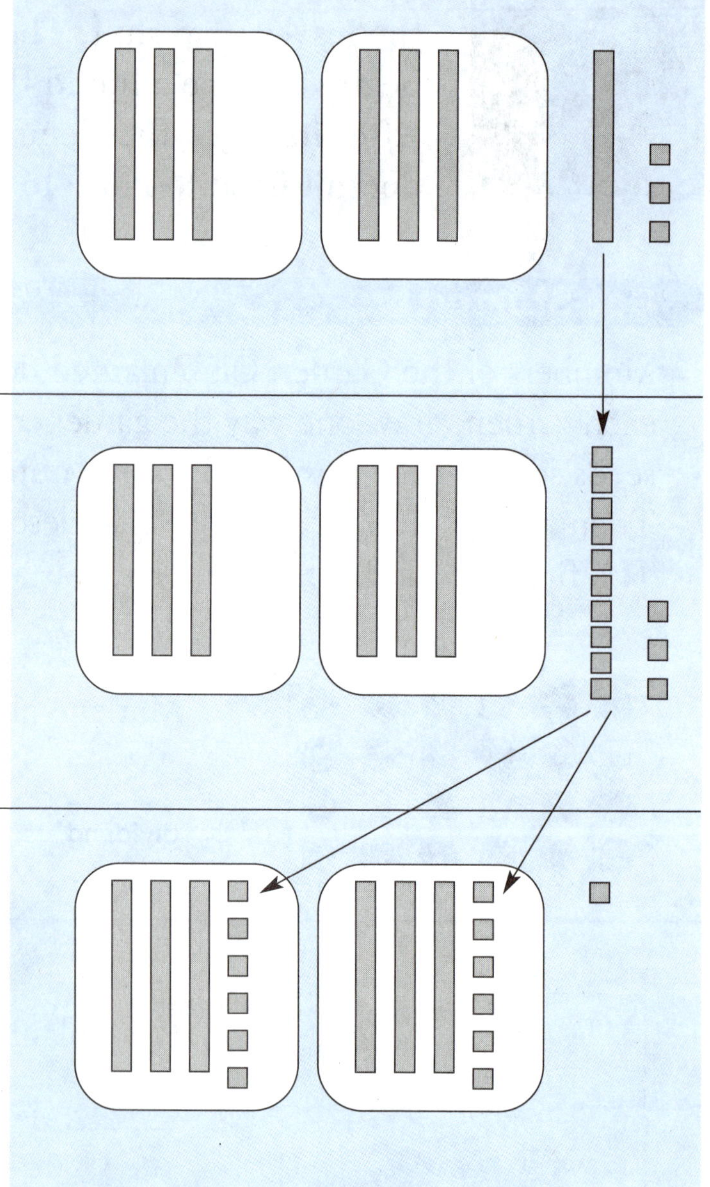

To divide 1 ten and 3 ones into 2 equal groups, start by regrouping the ten as 10 ones.

Now you have 13 ones.

Divide the 13 ones into 2 equal groups.

You can put 6 ones in each group.

You have 1 left over.

Each group has 36 and there is 1 left over. This leftover amount is called the **remainder**. A remainder of 1 is written as R1.

So 73 ÷ 2 = _______ R1

Use the drawings to help you find each quotient.

1. $47 \div 2 =$ _______

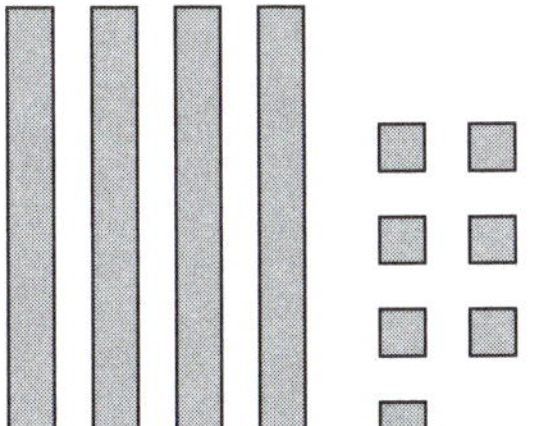 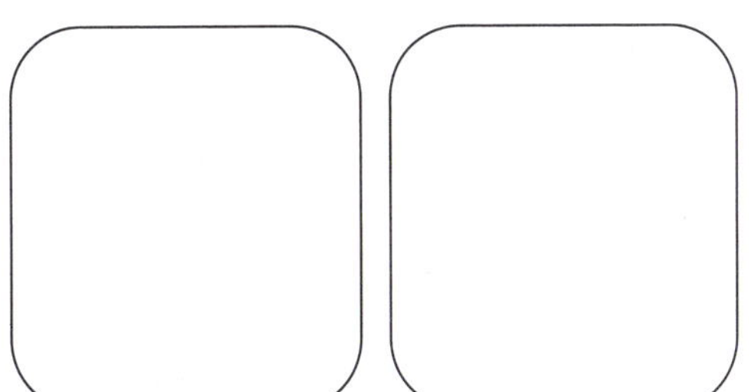

2. $52 \div 3 =$ _______

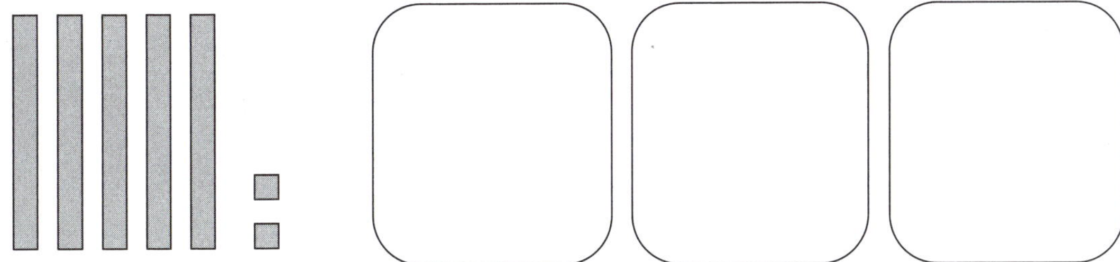

Solve a Problem

3. There are 65 students in Alissa's gym class. The teacher divided the students into 5 equal teams. How many students were on each team? Show your work.

There were _______ students on each team.

It's a Fact!

Another plant that eats insects is a Pitcher Plant. It is shaped like a small water pitcher.

Dividing without Drawings

Justin purchased a bag with 59 flower seeds.
If he shares all the seeds with the 5 members
of the Garden Club, how many seeds will each
member get? How many will be left over?

To solve this problem, you can find 59 ÷ 5.
Follow the steps below to divide without using a drawing.

STEP 1 Divide the 5 tens into 5 equal groups.
You can put 1 ten in each group.

Remember
Multiply and
subtract to find the
amount left over.

Multiply: $1 \times 5 = 5$.

Subtract: $5 - 5 = 0$.

You have 0 tens left over. ⟶

$$\begin{array}{r} 1 \\ 5\overline{)59} \\ -5 \\ \hline 0 \end{array}$$

STEP 2 Bring down the 9 ones.

Divide the 9 ones into 5 equal groups.
You can put 1 one in each group.

Multiply: $1 \times 5 = 5$.

Subtract: $9 - 5 = 4$.

You have 4 ones left over, ⟶
so the remainder is 4.

$$\begin{array}{r} 11 \\ 5\overline{)59} \\ -5 \\ \hline 09 \\ -5 \\ \hline 4 \end{array}$$

STEP 3 Write the quotient with the remainder.

59 ÷ 5 = _________ R_____

Each member of the Garden Club will get _________ seeds.

There will be _________ seeds left over.

Follow the steps you learned to find each quotient below.
Show your work.

1. $79 \div 7 =$ ___________

____

2. $85 \div 4 =$ ___________

____

3. $65 \div 3 =$ ___________

____

4. On a separate sheet of paper, find $73 \div 2$ by following the steps you learned. Then tell how the drawings on page 12 show the steps that you followed.

____

____

____

On Your Own

To help him check his answers to division problems, Greg wrote this rule:

(quotient $\times$ divisor) + remainder = dividend

Does Greg's rule always work? Explain.

1. $96 \div 4 = \square$

Ⓐ 16

Ⓑ 24

Ⓒ 25

Ⓓ 26

2. What is $48 \div 3$?

Ⓕ 15

Ⓖ 16

Ⓗ 15 R1

Ⓙ 16 R1

3. Kimiko and her class planted 75 flowers in the school's garden. They planted the flowers in 5 equal groups around the garden. How many flowers were in each group?

Ⓐ 11

Ⓑ 13

Ⓒ 15

Ⓓ 25

4. Find the quotient: $57 \div 4 = ?$

____

5. Use the drawing below to find $78 \div 3$.

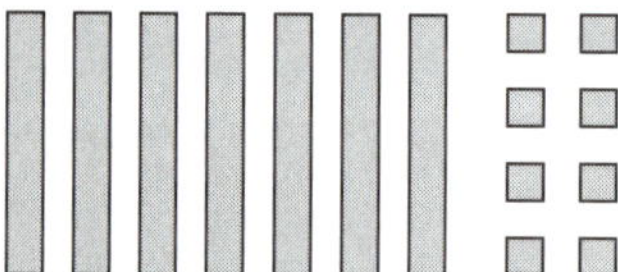

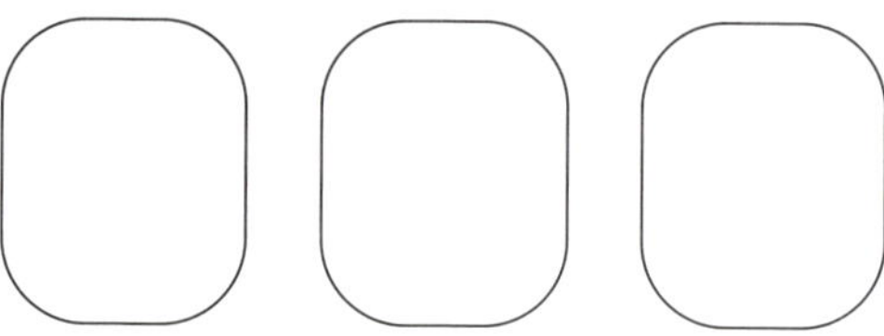

$78 \div 3 =$ _______

6. Think Back Find the product. Show your work.

$$\begin{array}{r} 207 \\ \times\ \ 8 \\ \hline \end{array}$$

Is It October Yet?

Since 1987, Americans have celebrated October as pizza month. Perhaps that's because pizza is a favorite American food. Each day, Americans eat about 100 acres of pizza. That's about 350 slices of pizza every second. Mushrooms and extra cheese are popular toppings. So are green peppers and onions. But what topping do Americans like best? It's pepperoni. Pizza lovers eat more than 250 million pounds of pepperoni each year.

Get Started

At Pizza Palace, there are three sizes of pizza. The large pizza has 8 slices. The medium pizza has 6 slices. The small pizza has 4 slices.

Write fractions to answer the questions below.

> **Remember**
> In a fraction, the bottom number tells how many equal parts are in the whole. The top number tells how many parts to count.

Large Pizza

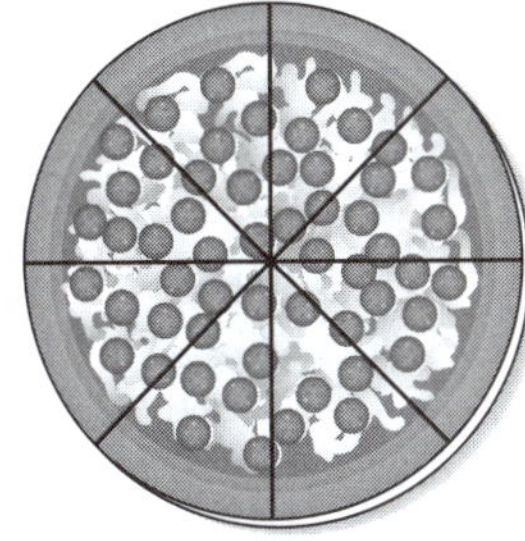

- What fraction represents the whole large pizza? _______________

Medium Pizza

- What fraction represents the whole medium pizza? _______________

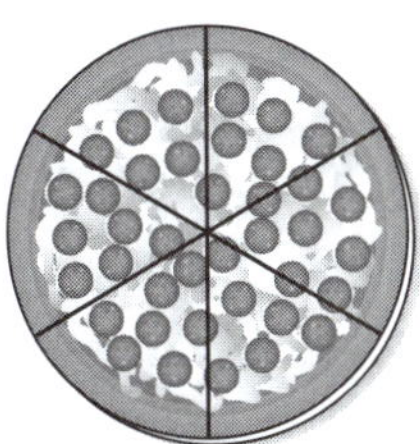

Small Pizza

- What fraction represents the whole small pizza? _______________

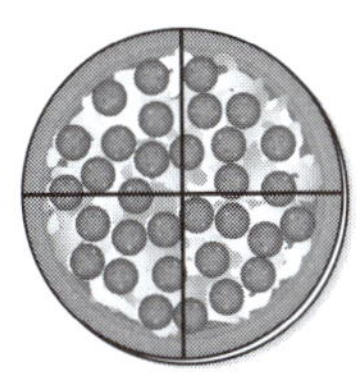

Working with Fractions and Mixed Numbers

Suppose you and a friend ordered 2 small pizzas at Pizza Palace. Together, you ate 6 slices. The shaded parts in the model below show the slices of pizza you and your friend ate.

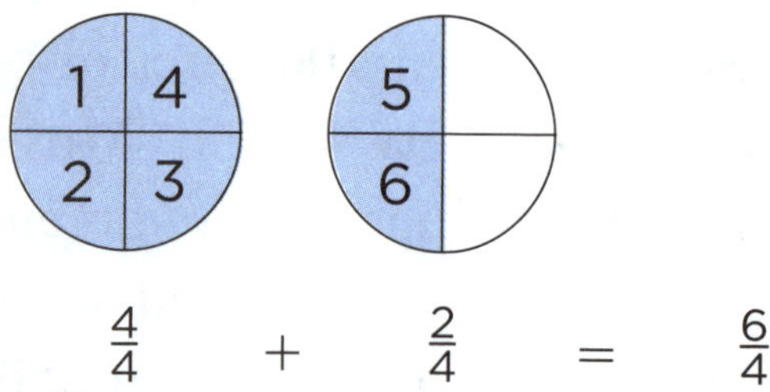

$$\frac{4}{4} \quad + \quad \frac{2}{4} \quad = \quad \frac{6}{4}$$

Look at the model above. Each shaded part represents one fourth of a pizza. Together, you and your friend ate six fourths, or $\frac{6}{4}$. The fraction $\frac{6}{4}$ is called an **improper fraction**.

You can also use a **mixed number** to represent the model. A mixed number has a whole number part and a fractional part. Look at the model below.

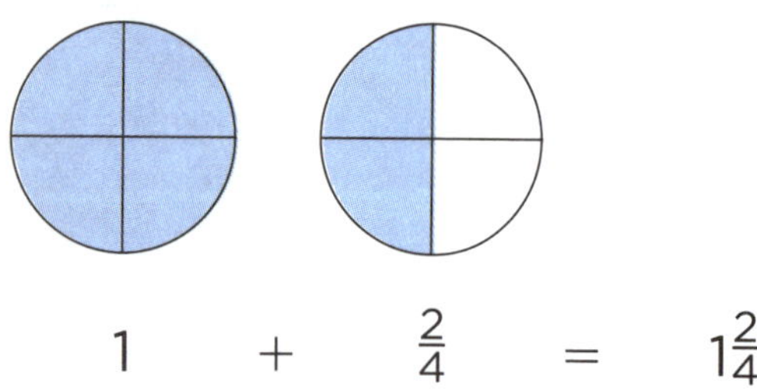

$$1 \quad + \quad \frac{2}{4} \quad = \quad 1\frac{2}{4}$$

There is 1 whole shaded and $\frac{2}{4}$ of another whole shaded.

So the model represents the mixed number $1\frac{2}{4}$.

1. As an improper fraction, $1\frac{2}{4}$ is written as _______.

Write an improper fraction and a mixed number to represent the shaded part of each model below.

2. 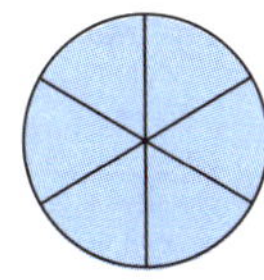 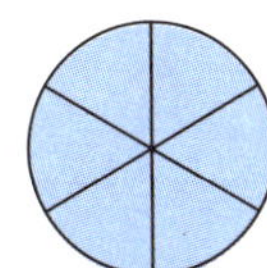 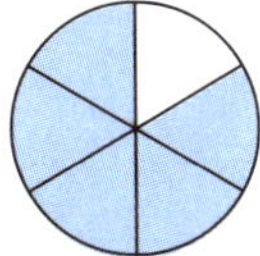

Improper Fraction: _________   Mixed Number: _________

---

3. 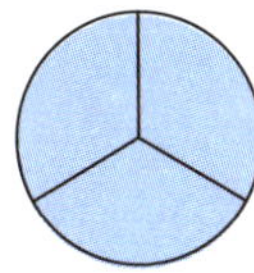 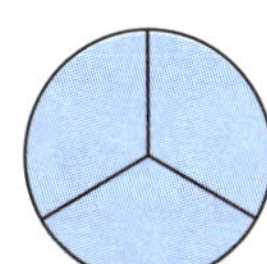 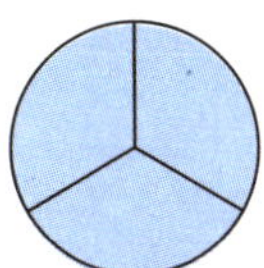 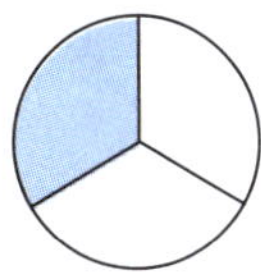

Improper Fraction: _________   Mixed Number: _________

---

4. 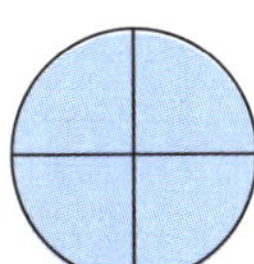 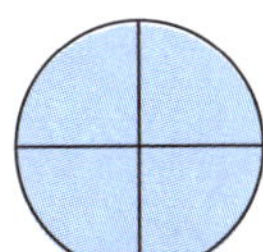 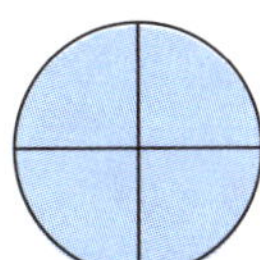 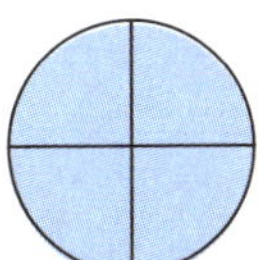 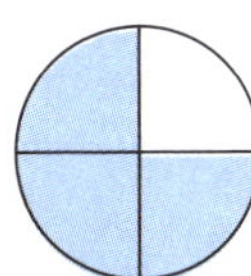

Improper Fraction: _________   Mixed Number: _________

Solve a Problem

5. If each student is served half an orange for lunch, how many oranges will be needed to serve 9 students? Show your work.

____

____

It's a Fact!

The largest pizza ever was made in South Africa in 1990. It measured 123 feet across at its widest part.

Improper Fractions and Mixed Numbers

To write an improper fraction, like $\frac{7}{3}$, as a mixed number,
you can use what you know about division. Follow the
steps below.

STEP 1 Divide the numerator by
the denominator.

> **Think:**
> $\frac{7}{3}$ means 7 divided
> into 3 equal groups.

$$3\overline{)7} \\ 2 \\ -6 \\ \overline{1}$$

STEP 2 In each group, there are
2 wholes and $\frac{1}{3}$ of another
whole.

So $\frac{7}{3} = 2\frac{1}{3}$.

To write the mixed number $2\frac{1}{3}$ as an improper fraction, you
can use multiplication and addition. Follow the steps below.

STEP 1 Multiply the denominator by the
whole number.

There are 2 wholes with 3 equal
parts in each whole: $2 \times 3 = 6$.

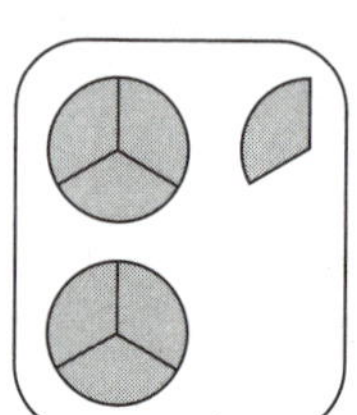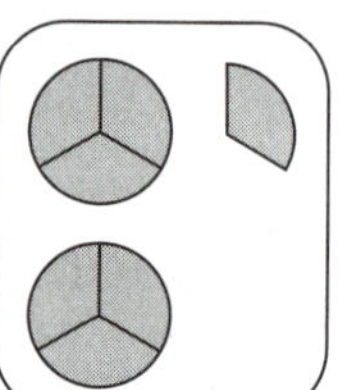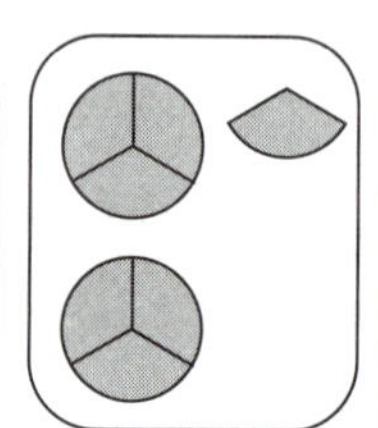

$$2\frac{1}{3}$$

STEP 2 Add the numerator to count the
extra third: $6 + 1 = 7$.

Write the sum over the denominator.

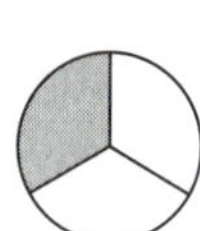

$$2\frac{1}{3} = \frac{7}{3}$$

Use the steps you learned to write the improper
fractions below as mixed numbers. Show your work.

1. $\frac{11}{4}$

2. $\frac{12}{5}$

3. $\frac{9}{2}$

Using the steps you learned, write the mixed numbers
below as improper fractions. Show your work.

4. $3\frac{2}{3}$

5. $1\frac{7}{8}$

6. $4\frac{1}{2}$

On Your Own

Draw a model to show
that $3\frac{1}{4} = \frac{13}{4}$.

1. Which of these is written as a mixed number?

- Ⓐ $\frac{2}{3}$
- Ⓑ $1\frac{3}{4}$
- Ⓒ 23
- Ⓓ $\frac{12}{5}$

2. Which of these numbers does the model below represent?

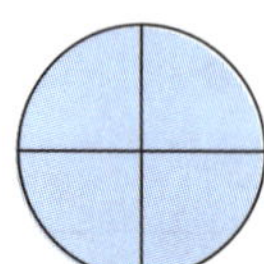 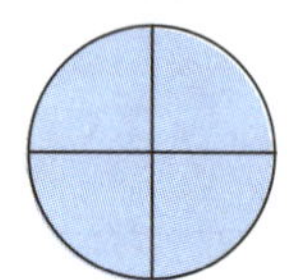 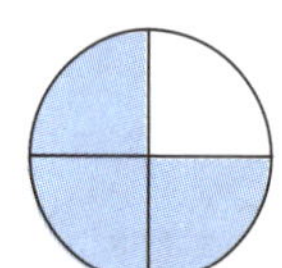

- Ⓕ $2\frac{1}{4}$
- Ⓖ $2\frac{3}{4}$
- Ⓗ $3\frac{3}{4}$
- Ⓙ 11

3. Which shows $1\frac{2}{3}$ written as an improper fraction?

- Ⓐ $\frac{12}{3}$
- Ⓑ $\frac{4}{3}$
- Ⓒ $\frac{5}{3}$
- Ⓓ $\frac{5}{2}$

4. Write $\frac{9}{8}$ as a mixed number. Show your work below.

____

5. Write the mixed number $2\frac{3}{5}$ as an improper fraction. Show your work below.

____

6. Think Back Find the quotient.

$$61 \div 5 = \underline{\hspace{2cm}}$$

Happy Birthday, U.S.A.

In 1776, church bells rang all over Philadelphia. The bells announced that representatives of the thirteen American colonies had approved a Declaration of Independence from Great Britain. Then on August 2, it became official. Someone, probably Timothy Matlack, wrote the Declaration by hand on a large sheet of parchment for members of the Continental Congress to sign. The first signature was John Hancock's, because he was President of the Congress. August 2 may be America's official birthday, but July 4th is the birthday Americans celebrate.

Get Started

Necie's mother made two desserts for a neighborhood Independence Day party. The models below represent the desserts. The shaded part of each model shows how much people ate.

- What fraction of the cake was eaten? _________

- What fraction of the cake was **not** eaten? _________

- Which of these fractions is greater? _________

Cake

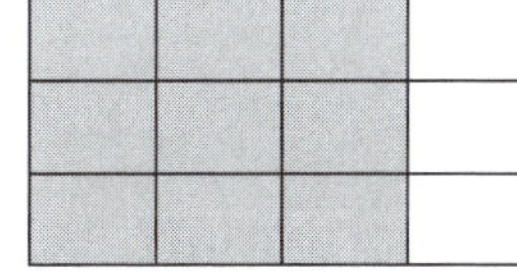

- What fraction of the pie was eaten? _________

- What fraction of the pie was **not** eaten? _________

- Which of these fractions is greater? _________

Pie

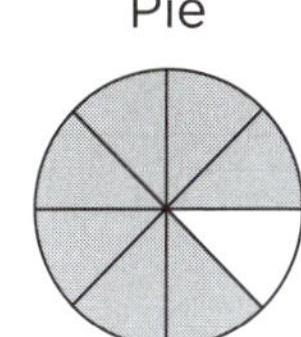

Comparing Fractions

To compare fractions that have the same **denominator,** like $\frac{3}{8}$ and $\frac{5}{8}$, you can compare the **numerators.**

You can use symbols to compare these fractions.

- $\frac{3}{8} < \frac{5}{8}$, because $3 < 5$.

- $\frac{5}{8} > \frac{3}{8}$, because $5 > 3$.

When you compare fractions that have different denominators, first find **equivalent fractions,** so that all of the fractions have the same denominator.

Equivalent fractions are fractions that name the same part of a whole. For example, $\frac{1}{2}$ and $\frac{2}{4}$ are equivalent fractions.

To find an equivalent fraction, you can multiply the numerator and denominator of the fraction by the same number.

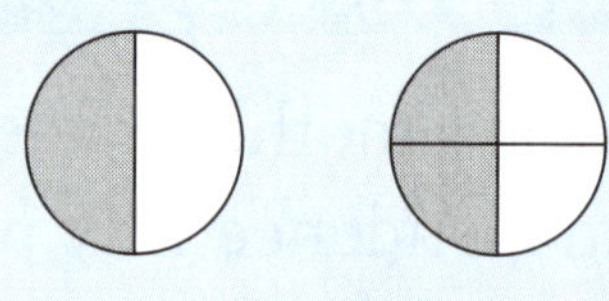

$$\frac{1}{2} \overset{\times 2}{\underset{\times 2}{=}} \frac{2}{4}$$

To compare $\frac{2}{3}$ and $\frac{5}{6}$, first find an equivalent fraction for $\frac{2}{3}$ with the same denominator as $\frac{5}{6}$.

Then compare the numerators: $\frac{4}{6} < \frac{5}{6}$, because $4 < 5$.

$$\frac{2}{3} \overset{\times 2}{\underset{\times 2}{=}} \frac{4}{6}$$

---

1. Write $>$, $<$, or $=$ to compare the fractions: $\quad \frac{2}{3} \bigcirc \frac{5}{6}$

Write >, < or = to compare the fractions in each pair.
Show your work.

2. $\dfrac{2}{3}$ ◯ $\dfrac{4}{9}$

3. $\dfrac{1}{2}$ ◯ $\dfrac{5}{10}$

4. $\dfrac{1}{2}$ ◯ $\dfrac{5}{6}$

5. $\dfrac{4}{5}$ ◯ $\dfrac{9}{10}$

6. $\dfrac{3}{8}$ ◯ $\dfrac{1}{4}$

7. $\dfrac{4}{10}$ ◯ $\dfrac{3}{5}$

Solve a Problem

8. A cake recipe calls for $\dfrac{5}{8}$ of a stick of butter.
Another cake recipe calls for $\dfrac{3}{4}$ of a stick of
butter. Which amount is greater, $\dfrac{5}{8}$ or $\dfrac{3}{4}$?
Show your work.

__________ is greater.

It's a Fact!

John Hancock signed his name very large so the King of England could read it without his glasses.

Ordering Fractions

If you know how to compare two fractions, you can also order three or more fractions.

Suppose you want to write the fractions $\frac{2}{3}$, $\frac{5}{6}$, and $\frac{9}{12}$ in order from least to greatest. Follow the steps below.

STEP 1 Write the fractions so they have the same denominator.

Find equivalent fractions for $\frac{2}{3}$ and $\frac{5}{6}$ with the same denominator as $\frac{9}{12}$.

$$\frac{2}{3} = \frac{8}{12} \quad (\times 4) \qquad \frac{5}{6} = \frac{10}{12} \quad (\times 2)$$

STEP 2 Compare the numerators. Order the fractions from least to greatest.

$$\frac{8}{12} \qquad \frac{9}{12} \qquad \frac{10}{12}$$

$$\frac{2}{3} \qquad \frac{9}{12} \qquad \frac{5}{6}$$

Use the $<$ symbol to write the fractions in order from least to greatest:

________ $<$ ________ $<$ ________

Use the < symbol to write the fractions in each group in order from least to greatest. Show your work.

1. $\frac{3}{4}, \frac{1}{2}, \frac{3}{8}$

_____ < _____ < _____

2. $\frac{5}{6}, \frac{1}{6}, \frac{3}{6}$

_____ < _____ < _____

3. $\frac{2}{3}, \frac{5}{12}, \frac{7}{12}$

_____ < _____ < _____

4. $\frac{7}{10}, \frac{3}{5}, \frac{9}{10}, \frac{4}{5}$

_____ < _____ < _____ < _____

On Your Own

Write $\frac{1}{2}$, $\frac{1}{4}$, and $\frac{1}{3}$ in order from least to greatest.

④ Test Yourself

1. Which is equivalent to $\frac{1}{2}$?

   Ⓐ $\frac{3}{10}$

   Ⓑ $\frac{4}{10}$

   Ⓒ $\frac{5}{10}$

   Ⓓ $\frac{8}{10}$

2. Which of these fractions is the least?

   Ⓕ $\frac{1}{3}$

   Ⓖ $\frac{1}{4}$

   Ⓗ $\frac{1}{5}$

   Ⓙ $\frac{1}{6}$

3. Which fraction below is the greatest?

   Ⓐ $\frac{5}{6}$

   Ⓑ $\frac{7}{12}$

   Ⓒ $\frac{1}{2}$

   Ⓓ $\frac{3}{4}$

4. Write the fractions below in order from least to greatest.

$$\frac{2}{3}, \ \frac{5}{6}, \ \frac{1}{2}$$

_______ < _______ < _______

5. Write two fractions below that are equivalent to $\frac{2}{3}$.

Explain your answer.

____

____

____

6. Think Back Write the improper fraction $\frac{8}{3}$ as a mixed number.

$$\frac{8}{3} = \underline{\hspace{2cm}}$$

Race to the Finish

Imagine running 26.2 miles in a little more than two hours. Many marathon runners do. They run so far that they wear out their shoes. To train for and finish a marathon, most runners need about three pairs of shoes. That wasn't a problem for Abebe Bikila from Ethiopia. He was the first African to win an Olympic medal. He earned a gold medal in the 1960 Olympic marathon. And he ran the race barefoot!

Get Started

Janice belongs to the Race Runners track club. The members take part in many races. Some are as short as 100 meters, and some are as long as a marathon.

- Janice ran a race that was 10 km long. After running 3 km, what fractional part of the race had she run? _______________

- After running 7 km, what fractional part of the race had Janice run? _______________

- After running 9 km, what fractional part of the race did she **have left** to run? _______________

- What does the denominator mean in each fraction you wrote?

____

____

____

Working with Decimals through Tenths

When a fraction or mixed number has a denominator of 10, you can easily write it as a **decimal** to the tenths place.

A decimal is a number that has a **decimal point.** The tenths place is to the right of the decimal point.

Ones	.	Tenths

The model at the right shows the fraction $\frac{7}{10}$. There are 10 equal parts in the whole, and 7 equal parts have been shaded. Each shaded part is a tenth.

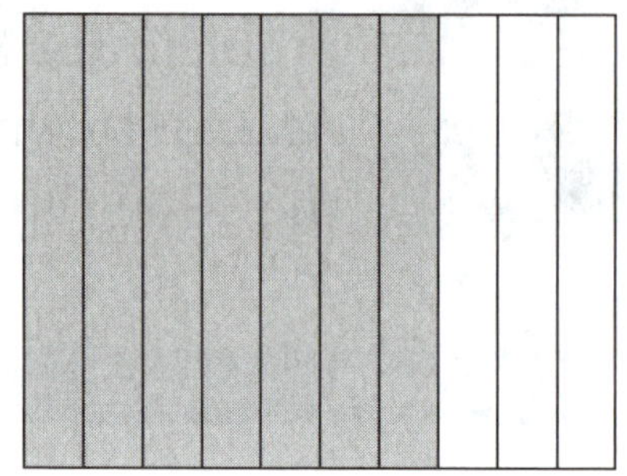

To write this fraction as a decimal, write the number of tenths shaded. There are 7 tenths shaded.

You can read 0.7 as "seven tenths."

Ones	.	Tenths
0	.	7

The model below shows the mixed number $2\frac{3}{10}$.

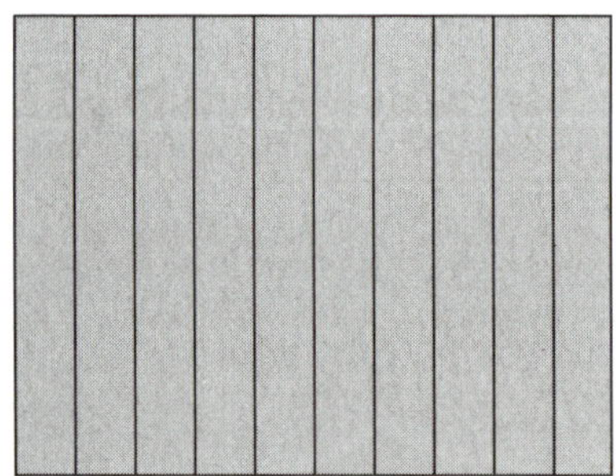  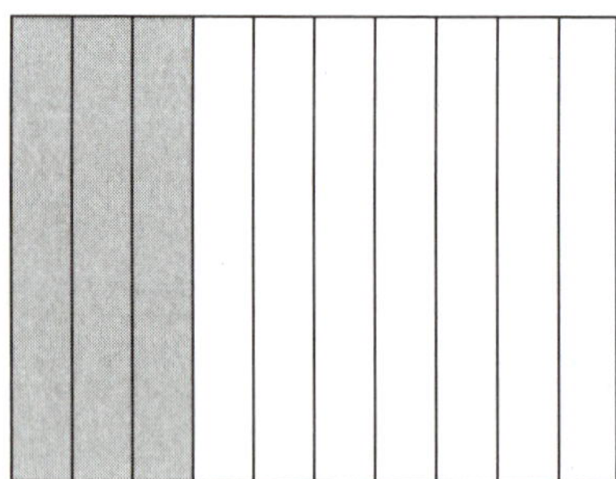

To write this mixed number as a decimal, first count the number of wholes shaded. This is the number of ones.

Ones	.	Tenths

Then count the number of shaded tenths that are left.

Write a fraction or mixed number with a denominator of 10 to represent the shaded part of each model below. Then write the fraction or mixed number as a decimal to the tenths place.

1. 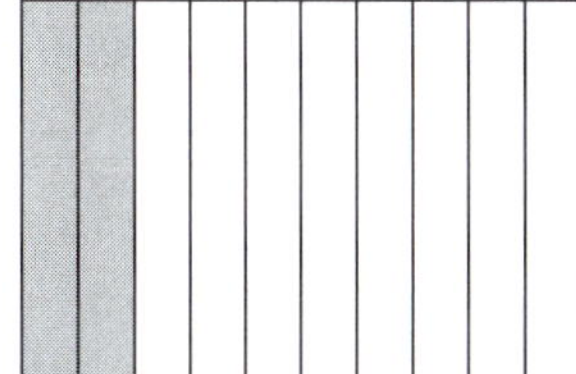

Fraction: _________

Decimal: _________

2. 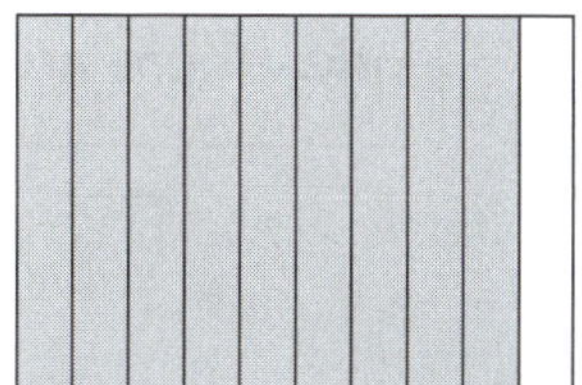

Mixed Number: _________

Decimal: _________

3. 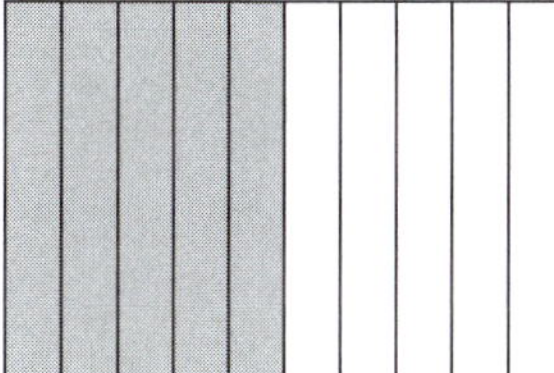

Fraction: _________

Decimal: _________

4. 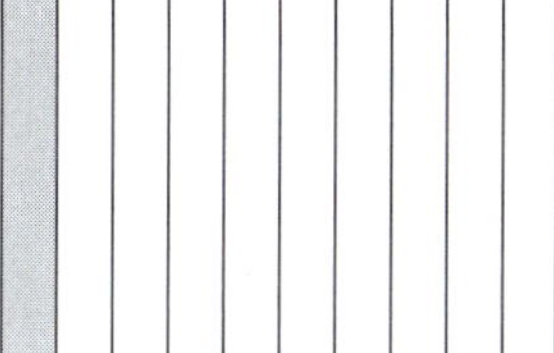

Mixed Number: _________

Decimal: _________

5. The deli manager sliced a block of cheese into ten equal pieces. He sold six pieces. What fraction of the cheese is left? Write your answer as a decimal.

____

It's a Fact!

The Boston Marathon is the oldest marathon that is run every year. It began in 1897.

Reading and Writing Decimals

You can use what you learned about decimals to help you read and write decimals. Look at the decimal below.

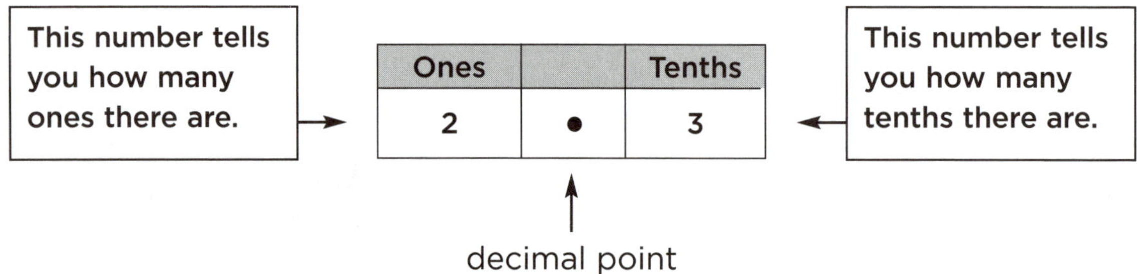

To read or write this decimal, follow the steps below.

STEP 1 Start with the whole number. Read or write it.

STEP 2 Read or write the decimal point as "and."

STEP 3 Read or write the digit to the right of the decimal point as a fraction.

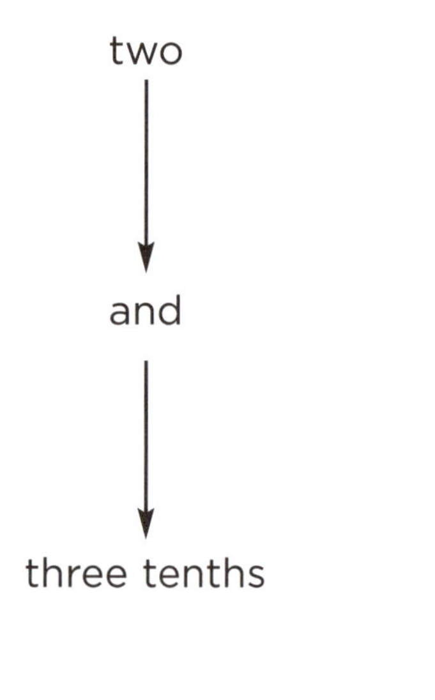

So, read or write 2.3 as "two and three tenths."

Follow the steps above to write 13.4 in words.

____

Use the steps you learned to write each of these decimals in words.

1. 9.7

____

____

2. 0.9

____

____

3. 15.2

____

____

4. 20.5

____

____

Write the decimal for each amount below.

5. three and eight tenths

____

6. one tenth

____

7. In the decimal 0.8, what does the "0" in the ones place mean? Explain your answer on the lines below.

____

____

On Your Own

Write each of the numbers below as a decimal and in words:

$$\frac{6}{10} \qquad 3\frac{1}{10}$$

$$1\frac{5}{10} \qquad \frac{4}{10}$$

1. Diana has completed three tenths of her homework. Which decimal names this amount?

Ⓐ 0.3

Ⓑ 10.3

Ⓒ 3.10

Ⓓ 31.0

2. Bob ran 0.7 of a mile. What fraction of a mile did he run?

Ⓕ $\frac{1}{10}$

Ⓖ $\frac{1}{7}$

Ⓗ $\frac{0}{7}$

Ⓙ $\frac{7}{10}$

3. Which shows another way to write 3.9?

Ⓐ 39

Ⓑ 0.39

Ⓒ $3\frac{9}{10}$

Ⓓ $3\frac{9}{100}$

4. Write a fraction and a decimal to represent the shaded amount below.

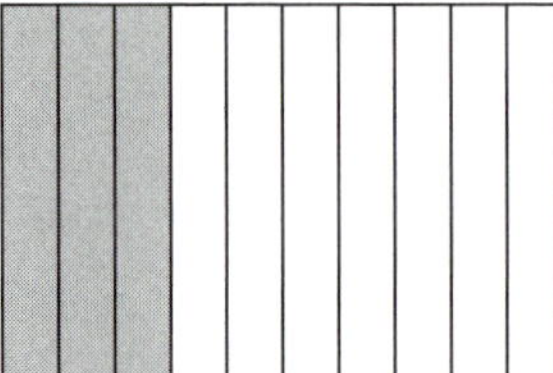

Fraction: _________

Decimal: _________

5. Write the decimal 9.7 in words.

____

6. Think Back   Write the fractions below in order from least to greatest.

$$\frac{5}{6} \qquad \frac{2}{3} \qquad \frac{1}{2}$$

_______ < _______ < _______

The World's Oldest Money

Did you know that the world's oldest form of money is a seashell? The shell belongs to an animal called a cowrie (COW ree). Cowries are mollusks, the group of water animals that include clams and oysters. They live in shallow water in the Pacific and Indian Oceans. Thousands of years ago, people in China began using cowrie shells as money. The practice spread around the world, lasting into the 1900s in parts of Africa.

Get Started

A group of students were going to the store to buy new notebooks. Sarah had 4 dimes and 9 pennies. Mark had 5 dimes and 8 pennies. Ted had 4 dimes and 7 pennies. Suki had 7 dimes and 4 pennies.

You can use the chart below to write the amount of money each student had as a decimal. For each student, write the number of dimes and pennies.

			Dimes	Pennies
Sarah $	0	.		
Mark $	0	.		
Ted $	0	.		
Suki $	0	.		

- Which student had $0.47? _________________

- Which student had $0.74? _________________

- Which student had the greatest amount? _________________

Working with Decimals through Hundredths

When a fraction or mixed number has a denominator of 100, you can easily write it as a decimal to the hundredths place.

In a decimal, the hundredths place is to the right of the tenths place:

Ones	.	Tenths	Hundredths

The model below shows the fraction $\frac{35}{100}$. There are 100 equal parts in the whole, and 35 equal parts are shaded.

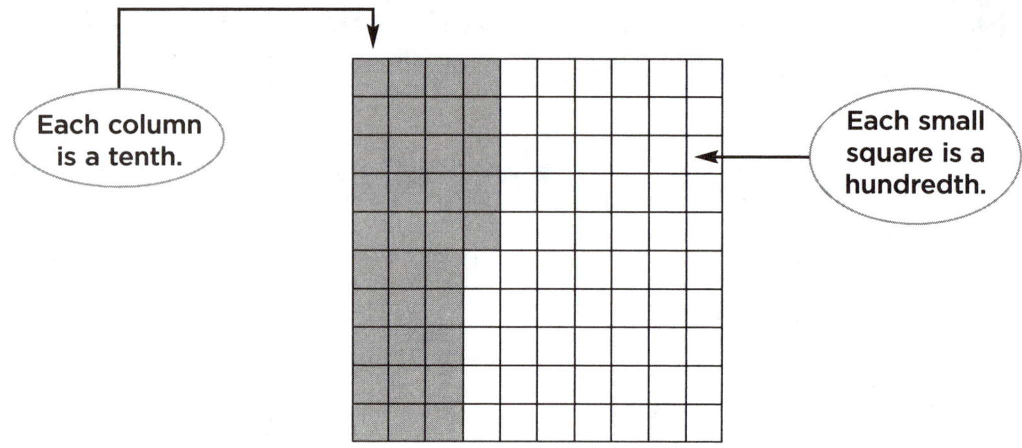

To write this fraction as a decimal, first count the number of tenths shaded. There are 3 tenths shaded.

Ones	.	Tenths	Hundredths
0	.	3	5

Then count the shaded hundredths that are left. There are 5 shaded hundredths left.

As a decimal, $\frac{35}{100}$ can be written as 0.35. Read this decimal as "thirty-five hundredths."

1. In the decimal 0.35, what does the 0 mean?

____

____

Write a fraction with a denominator of 100 to represent the shaded part of each model below. Then write the fraction as a decimal to the hundredths place.

2.

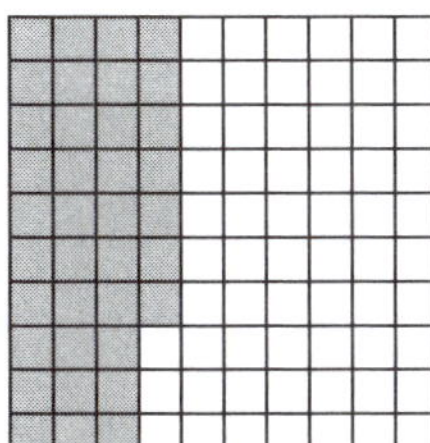

Fraction: ______________

Decimal: ______________

3.

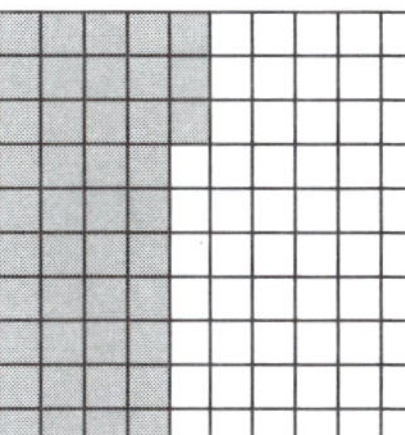

Fraction: ______________

Decimal: ______________

4.

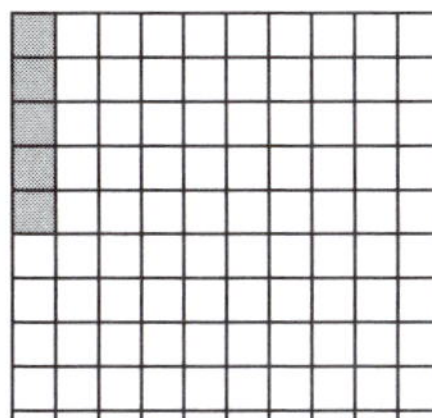

Fraction: ______________

Decimal: ______________

5.

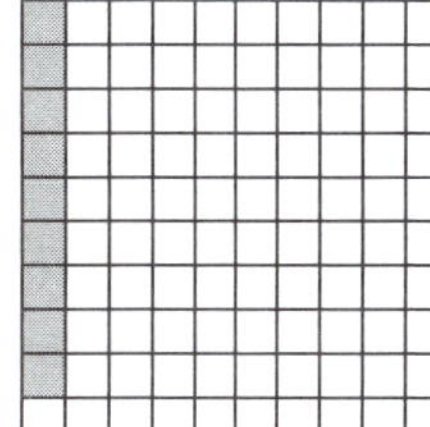

Fraction: ______________

Decimal: ______________

Solve a Problem

6. Write a fraction and a decimal to represent the shaded part of the model below.

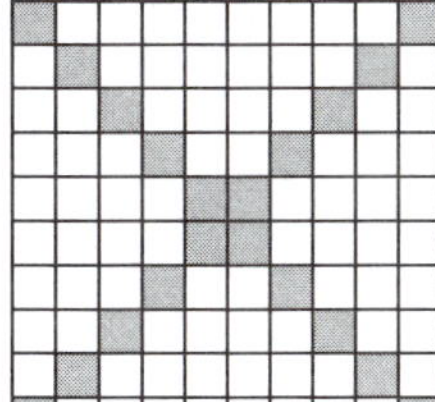

Fraction: ______________

Decimal: ______________

It's a Fact!

Some people think the phrase "shell out the money" comes from the time when cowries were used as money.

Equivalent Fractions and Decimals

You know how to write a fraction as a decimal when the fraction has a denominator of 10 or 100. But what if the fraction has a different denominator?

Suppose you want to write the fraction $\frac{3}{4}$ as a decimal.

Follow the steps below.

STEP 1 Write an equivalent fraction with a denominator of 10 or 100.

$$\overset{\times 25}{\frac{3}{4}} = \underset{\times 25}{\frac{75}{100}}$$

Remember
To find an equivalent fraction, you can multiply the numerator and denominator by the same number.

STEP 2 Write this fraction as a decimal to the tenths or hundredths place.

You can use a model to help you.

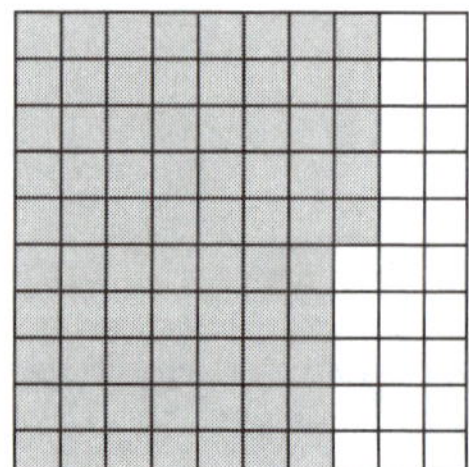

$$\frac{75}{100} =$$

Ones	.	Tenths	Hundredths
0	.		

Use the steps above to write $\frac{1}{2}$ as a decimal to the tenths place.

$$\frac{1}{2} =$$

Ones	.	Tenths
0	.	

Use the steps you learned to write each fraction as a decimal to the tenths or hundredths place. Show your work.

1. $\dfrac{1}{5}$ =

Ones	.	Tenths
0	.	

2. $\dfrac{3}{50}$ =

Ones	.	Tenths	Hundredths
0	.		

3. $\dfrac{3}{5}$ =

Ones	.	Tenths
0	.	

4. $\dfrac{8}{25}$ =

Ones	.	Tenths	Hundredths
0	.		

5. $\dfrac{9}{20}$ =

Ones	.	Tenths	Hundredths
0	.		

On Your Own

Use what you have learned to write the mixed number $2\frac{1}{4}$ as a decimal. Draw a model and show all of your work.

1. Which shows $\frac{65}{100}$ written as a decimal?

    (A) 0.65

    (B) 6.5

    (C) 0.065

    (D) 65

2. Roberto ran 0.49 mile. What part of a mile did he run?

    (F) 4.9

    (G) 0.049

    (H) $\frac{49}{10}$

    (J) $\frac{49}{100}$

3. Which decimal below represents this model?

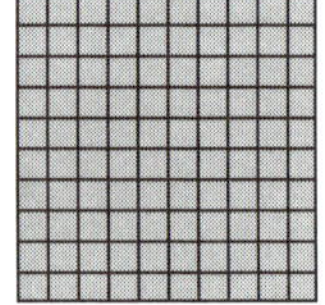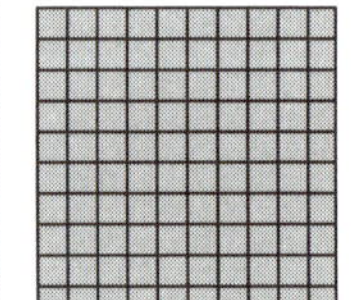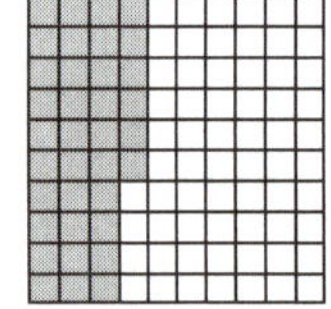

    (A) 32.6

    (B) 3.26

    (C) 2.36

    (D) 0.326

4. Write the fraction $\frac{3}{5}$ as a decimal.

____

5. Shade in the model below to show the fraction $1\frac{13}{20}$.

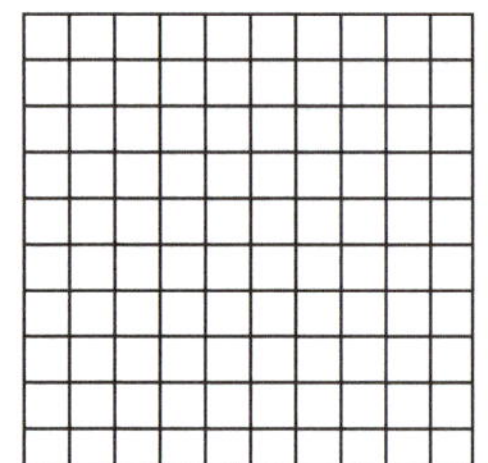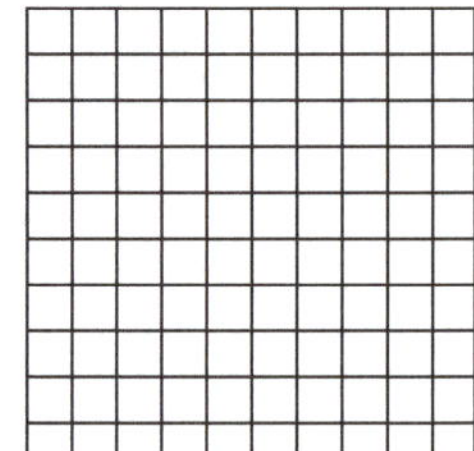

Write $1\frac{13}{20}$ as a decimal.

____

6. **Think Back** Find the product. Show your work.

$$128 \times 3$$

No Ordinary Mirror

How hard is it to haul a mirror? If the mirror is 27 feet wide and weighs 18 tons, it takes many people, careful planning, and a special truck. The truck carried its load one mile per hour up a winding mountain road in Arizona. There, the mirror became part of the world's largest reflecting telescope. The telescope, in a building 16 stories tall, will help scientists see galaxies far beyond the Milky Way.

Get Started

Students in Tyler's class each drew a dotted line on a piece of paper. Then they drew half of a figure on one side of the dotted line. Next, they traded papers and asked someone in the class to complete the figure.

Tyler gave his figure to Carla. She put a mirror along the dotted line. Then she drew the reflection she saw in the mirror.

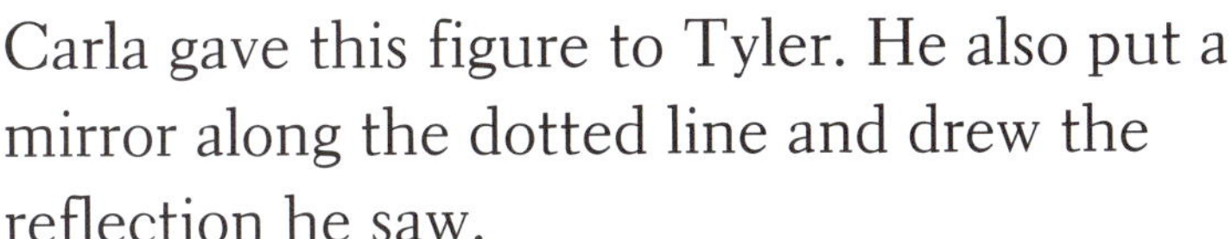

- What did Carla draw to complete the figure?

____

Carla gave this figure to Tyler. He also put a mirror along the dotted line and drew the reflection he saw.

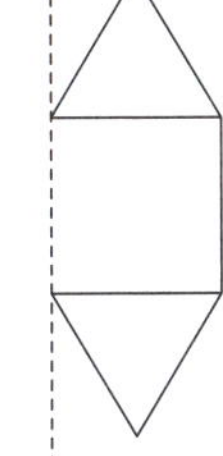

- What did Tyler draw to complete the figure?

____

- Complete the drawings above to show their reflections.

Working with Flips and Turns

In geometry, there are special ways to move a figure from one place to another without changing its shape or size.

One way is to turn a figure around a point. This is called a rotation.

To rotate a figure, first choose a point on the figure that will not move. This is called the point of turn.

Look at the pictures below.

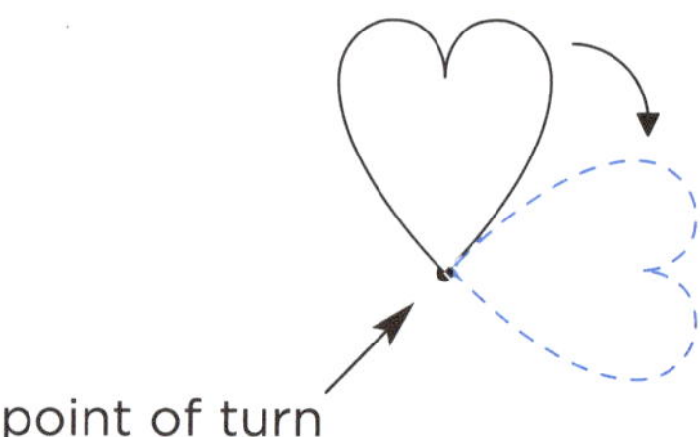

This shape has been rotated to the right, or **clockwise**.

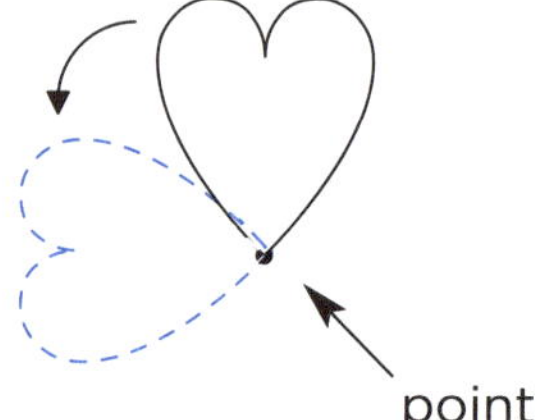

This shape has been rotated to the left, or **counterclockwise**.

Another way to move a figure without changing its shape or size is to flip the figure over a line. This is called a reflection.

Look at the pictures below.

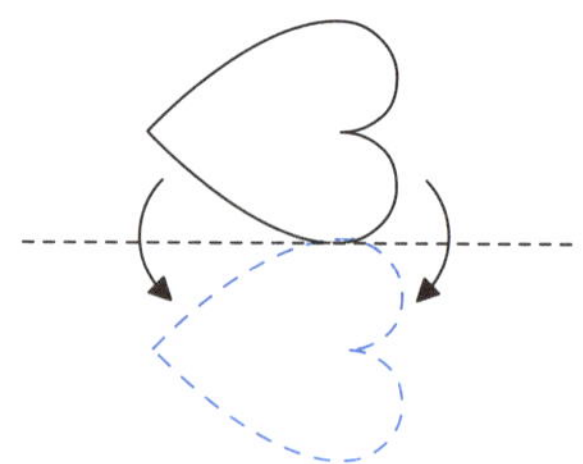

This shape has been reflected over a horizontal line.

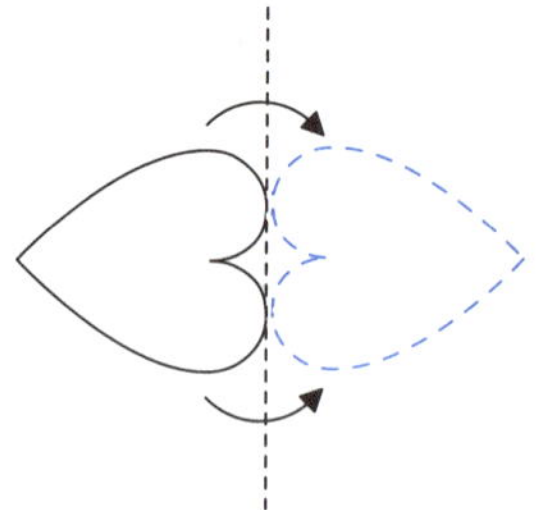

This shape has been reflected over a vertical line.

42   Level D

Tell whether the picture shows a rotation or a reflection. If it is a rotation, draw the point of turn. If it is a reflection, draw the horizontal or vertical line of reflection.

1.

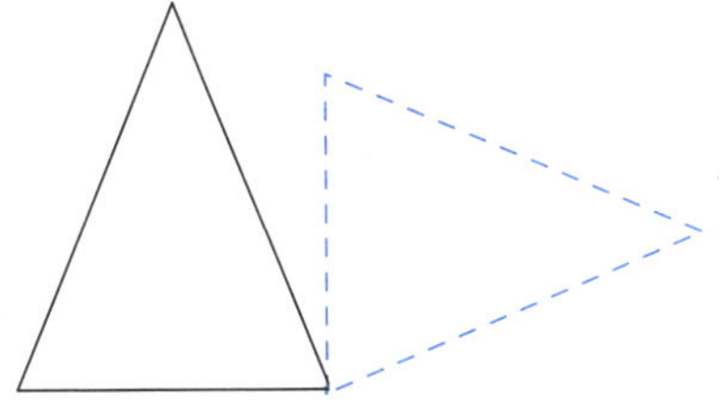

____

2.

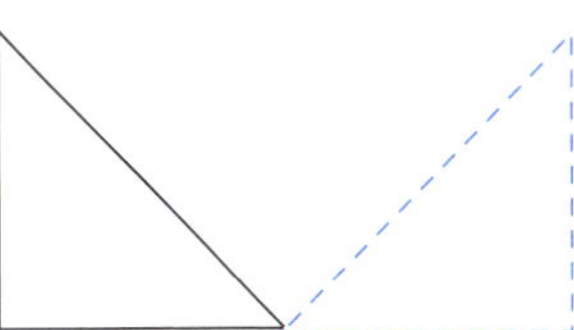

____

3.

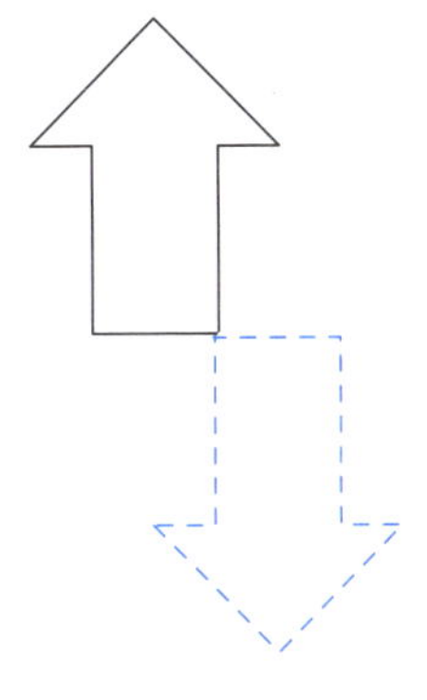

____

4.

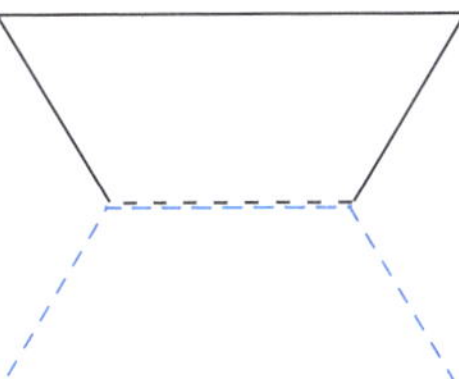

____

Solve a Problem

5. Look at the figures above. Are the rotations and reflections congruent to the original figures? Explain.

____

____

____

It's a Fact!

A special steel box was made to carry the LBT's mirror. Together, the mirror and its box weighed 55 tons. That's equal to the weight of about 7 African elephants.

Showing Rotations and Reflections

You can use what you have learned to show your own rotations and reflections.

Start by tracing or drawing a shape or design on a sheet of paper. Cut out your drawing. Then follow the steps below.

Follow these steps to show a rotation.

STEP 1 Choose any point on the cutout drawing to be the point of turn.

Use your finger or a pencil to hold down the figure at this point.

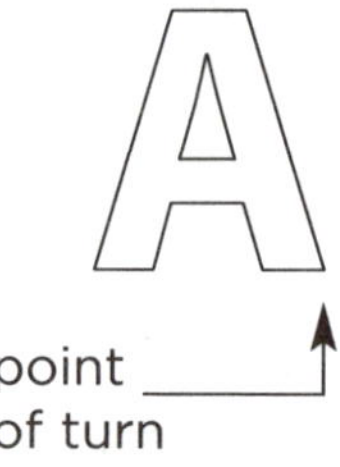

STEP 2 Rotate the figure clockwise or counterclockwise. Do not move the point of turn.

Then trace the rotated figure and show the point of turn.

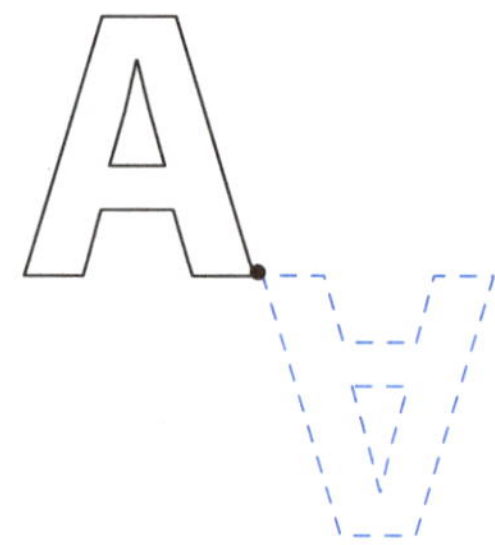

Follow these steps to show a reflection.

STEP 1 Draw a horizontal or vertical line next to your cutout drawing as shown.

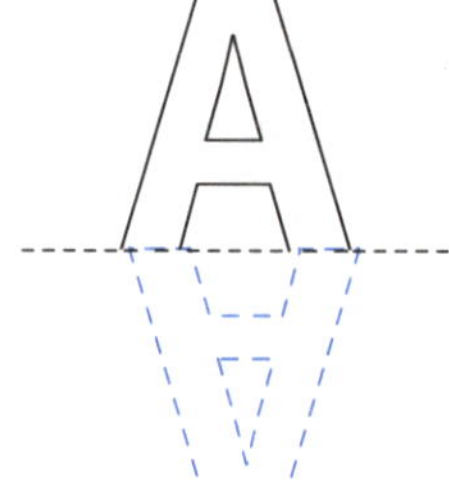

STEP 2 Flip the figure over the line and trace the reflection.

Which line is used in the picture above—
a horizontal line or a vertical line? _______________

Trace each of the figures below. Cut out the figures.
Then follow the steps on page 44 to show a rotation
or reflection.

1. Show a reflection of this figure over a vertical line.

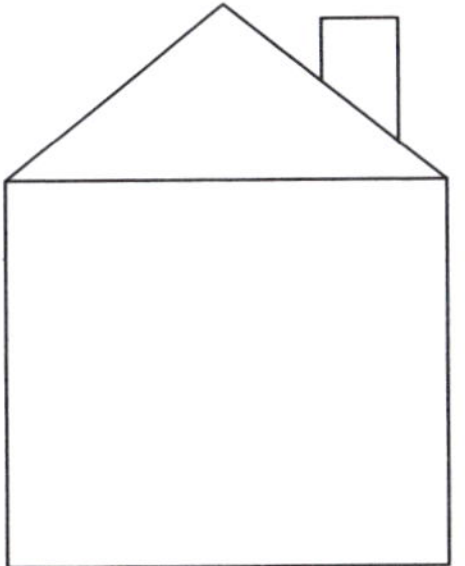

2. Show a clockwise rotation of this figure.

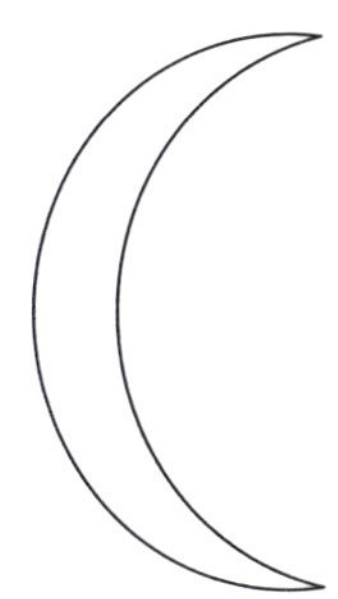

3. Draw a reflection of this figure over the
horizontal line shown.

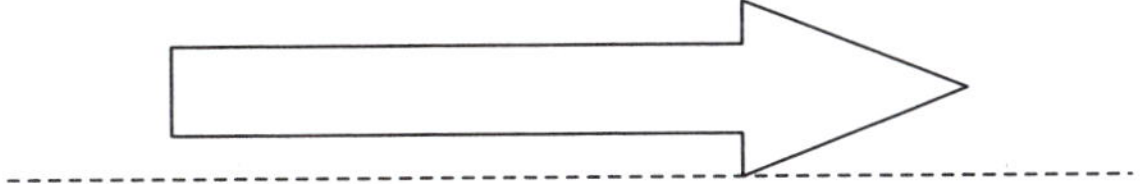

On Your Own

Use the digits 1–9 and create
flip designs. For example, with
6 you can make this design:

Try it!

1. Which statement about congruent figures is true?

  Ⓐ They are the same shape, but they are different sizes.

  Ⓑ They have a different shape and a different size.

  Ⓒ They have the same shape and the same size.

  Ⓓ They have a different shape, but they are the same size.

2. Which is another name for a flip?

  Ⓕ reflection

  Ⓖ rotation

  Ⓗ line of symmetry

  Ⓙ congruence

3. Which shows the letter Q reflected over a vertical line?

  Ⓐ

  Ⓑ

  Ⓒ

  Ⓓ

4. The point that does not move in a rotation is called the

____

5. Draw and rotate the letter R.

First, rotate the letter R clockwise. Draw the rotated figure here.

Now rotate the letter R counterclockwise. Draw the rotated figure here.

6. Think Back Show the lines of symmetry on the rectangle below.

How many lines of symmetry does the rectangle have?

____

America's Cup

In 1851, Queen Victoria invited the world's best sailors to England for a 60-mile race for a silver cup. The queen watched the race from aboard her royal yacht. People expected one of the 16 British ships in the race to win. But the schooner *America* was the first ship to dip its flag in salute to the queen. The U.S. was the winner in a race that came to be called *America's Cup.*

Get Started

Mr. Lauren and his crew set up the floating markers, or buoys, for the first day of a two-day sailing race.

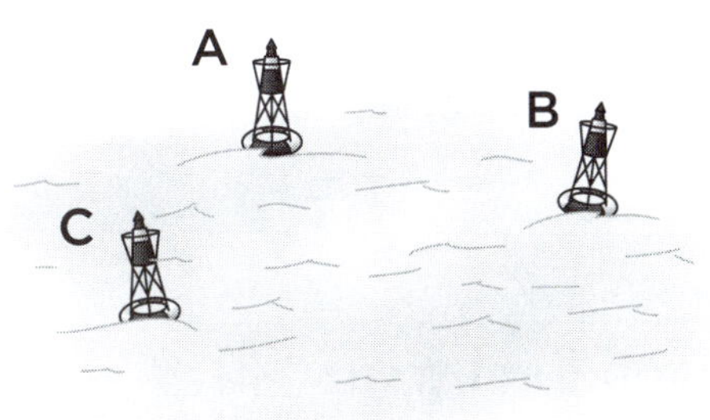

- Use your pencil and a ruler to connect the buoys in alphabetical order. Then connect Buoy C to Buoy A. What shape do the buoys make in the water?

____

On the second day, the crew added another buoy.

- Use your pencil and a ruler to connect the buoys in alphabetical order. Then connect Buoy D to Buoy A. What shape do the buoys make in the water?

____

Working with Geometric Figures

Geometric figures such as points, line segments, lines,
rays, and angles can help you talk about and model
objects in the world around you.

For example, you can use points to represent exact
locations, like the locations of Mr. Lauren's buoys:

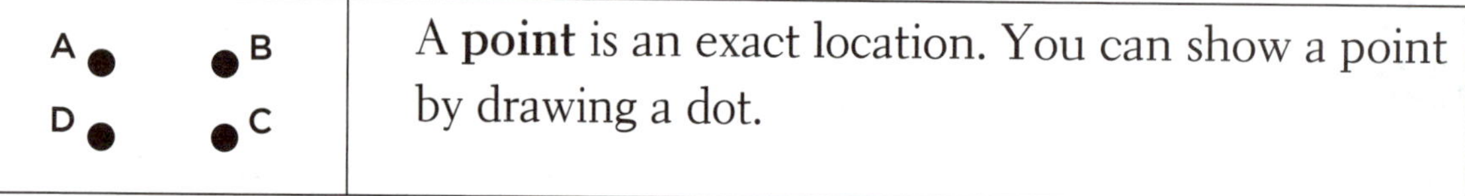

A **point** is an exact location. You can show a point
by drawing a dot.

You can use a line segment to represent the distance
between two points. The line segment below represents the
distance between two of the buoys:

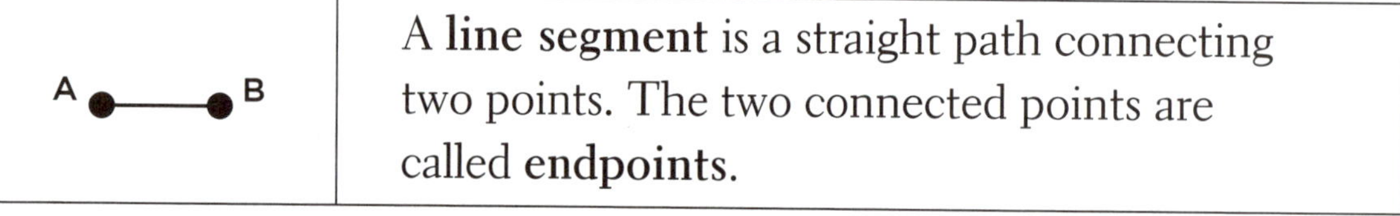

A **line segment** is a straight path connecting
two points. The two connected points are
called **endpoints**.

Here are some other geometric figures you can use:

A B	A **line** is a straight path that goes on forever in two directions. A line has no endpoints.
A B	A **ray** has one endpoint. It goes on forever in one direction.
A B D	An **angle** is formed by two rays that share the same endpoint. This endpoint is called the **vertex** of the angle. The vertex of this angle is point A.

1. Name the geometric figures you can
   see in the drawing at the right.

____

____

In the column on the left, write the name of each geometric figure shown. Then draw line segments to match each geometric figure with a picture in the right column.

2.

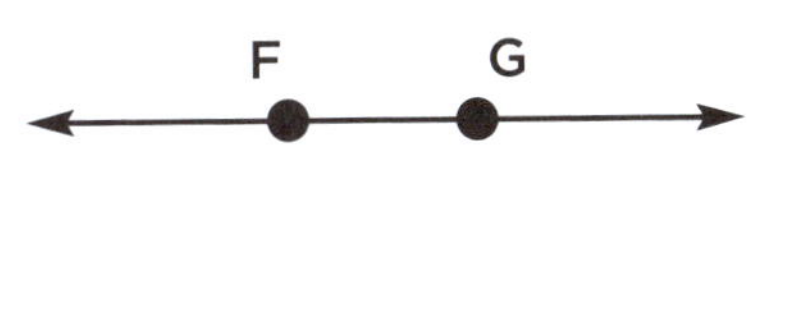

____

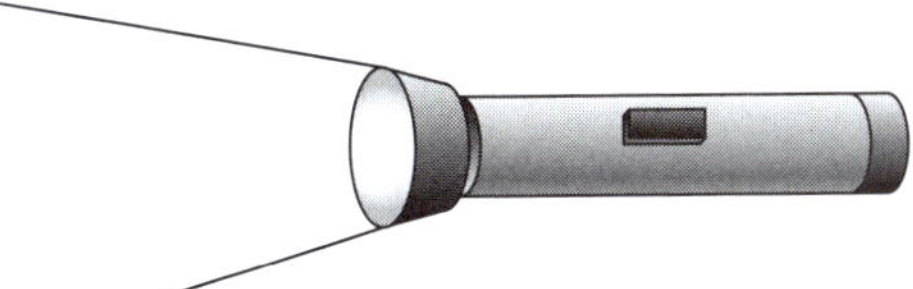

3.

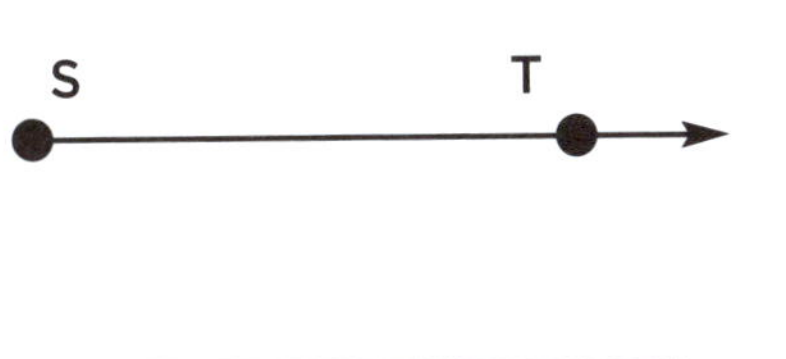

____

4.

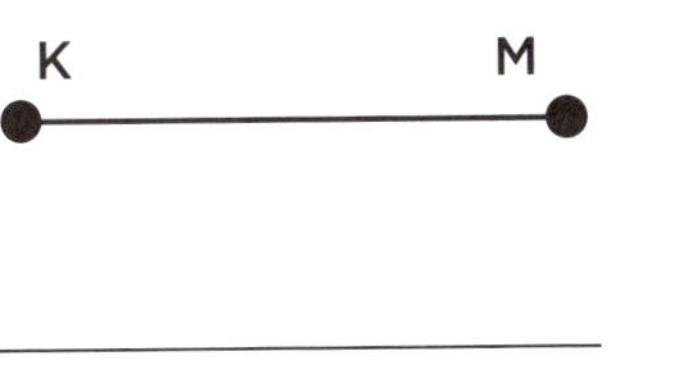

____

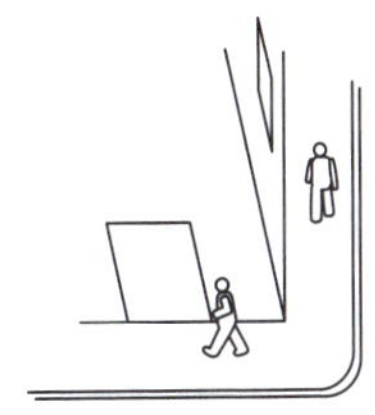

5.

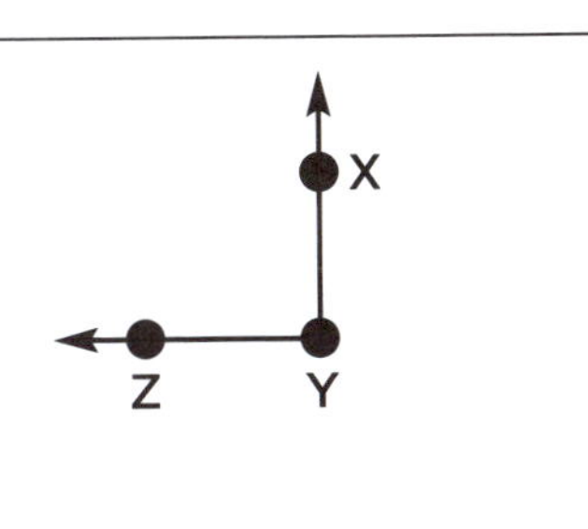

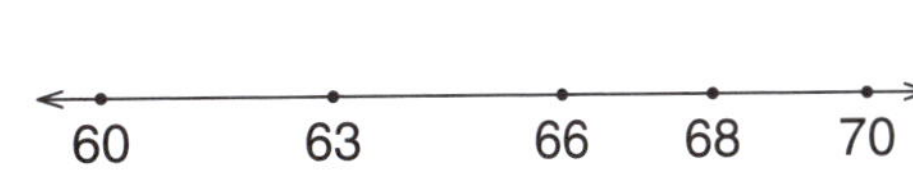

____

6. Use the points below to draw a ray.

 P    Q

It's a Fact!

Only three nations have ever won the America's Cup—the U.S., Australia, and New Zealand.

Naming Kinds of Angles

You learned that an angle is formed by two rays that share the same endpoint, which is called the **vertex** of the angle.

You can think of the hands on a clock as rays. They form many different angles during the day. Use the clock faces below and follow the steps to name three different kinds of angles.

STEP 1 Draw the hour hand on this clock to show 3:00.

The two rays form a **right angle**. A right angle has a square corner.

STEP 2 Draw the hour hand on this clock to show 4:00.

These two rays form an **obtuse angle**. An obtuse angle has a wider opening than a right angle.

STEP 3 Draw the hour hand on this clock to show 10:00.

These two rays form an **acute angle**. An acute angle has a narrower opening than a right angle.

Name the kind of angle formed by the hands of a clock at these times:

7:00 __________   9:00 __________   5:00 __________

Write *acute angle*, *obtuse angle*, or *right angle*
to describe each angle below.

1.

____

2.

____

3.

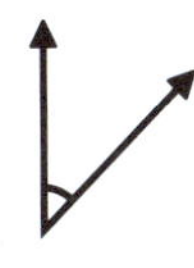

____

4.

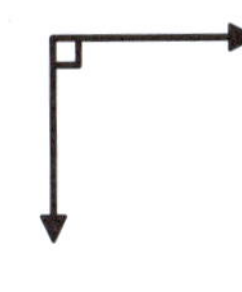

____

5.

____

6.

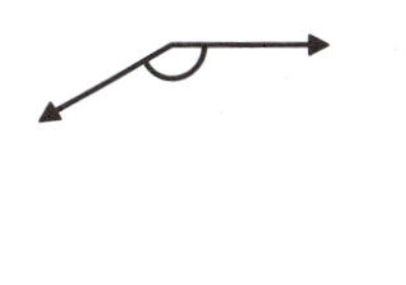

____

Complete each sentence with the correct word.

7. A(n) __________ is a straight path that goes on forever in two directions.

8. The endpoint shared by two rays that form an angle is called the __________.

9. A __________ goes on forever in one direction and has just one endpoint.

On Your Own

Look around your home and list five objects that have angles. Draw and name the angles.

8 Test Yourself

1. A part of a line that has only one endpoint is called a . . .

Ⓐ point

Ⓑ ray

Ⓒ line

Ⓓ line segment

2. What kind of angle does the inside of the letter L represent?

Ⓕ obtuse

Ⓖ straight

Ⓗ acute

Ⓙ right

3. The distance between two bases on a baseball diamond is an example of a ______________.

Ⓐ point

Ⓑ line

Ⓒ ray

Ⓓ line segment

4. What kind of angle is formed when the hands of a clock are at 11:00?

____

5. Draw and label an acute angle on this clock.

What time does your clock show?

____

6. Think Back Multiply. Show the partial products.

$$\begin{array}{r} 72 \\ \times\ 5 \\ \hline \end{array}$$

The Transcontinental Railroad

In 1862, President Abraham Lincoln signed an act to build a railroad to connect the nation. The government chose two companies to do the job. In 1863, the Central Pacific started in California and built eastward. The Union Pacific broke ground in Nebraska and built westward. One month before the last spike went in the ground, the companies agreed where to meet. The golden spike went in the ground at Promontory Summit, Utah, in 1869.

Get Started

Ari's class went on a field trip to a railroad museum. While he was there, Ari looked for objects that could represent geometric figures. He drew pictures of what he found.

Here are some of the objects Ari drew. Name the geometric figure each object represents.

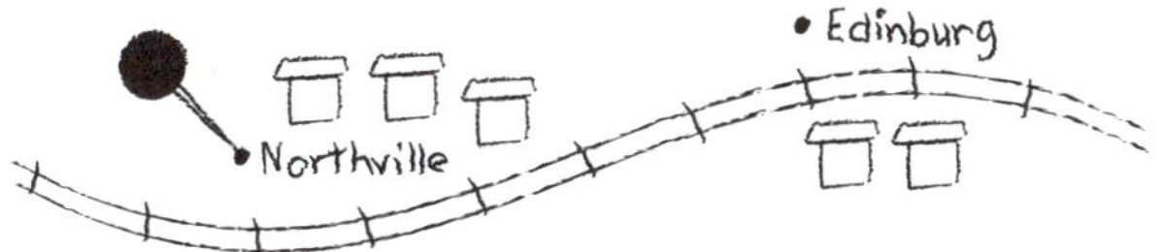

A map pin represents a(n) ___________.

Each side of a train's number plate represents a(n) ___________.

A beam of light from a headlight represents a(n) ___________.

Intersecting and Perpendicular Lines

You have learned that a line is a straight path that goes on forever in two directions.

When two or more lines cross, they are called **intersecting lines**.

Look at the picture at the right. You can think of the street and the railroad tracks as intersecting lines. The point at which they cross is called the **point of intersection**.

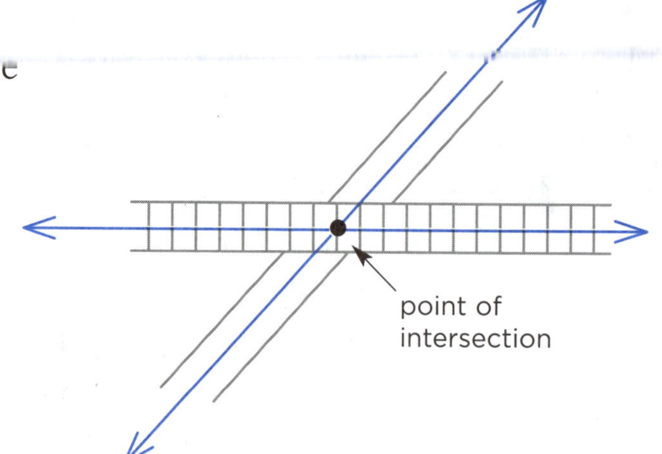

When intersecting lines form square corners, or right angles, they are called **perpendicular lines**.

The picture at the right shows a street and railroad tracks crossing as perpendicular lines. When two lines intersect as perpendicular lines, four right angles are formed.

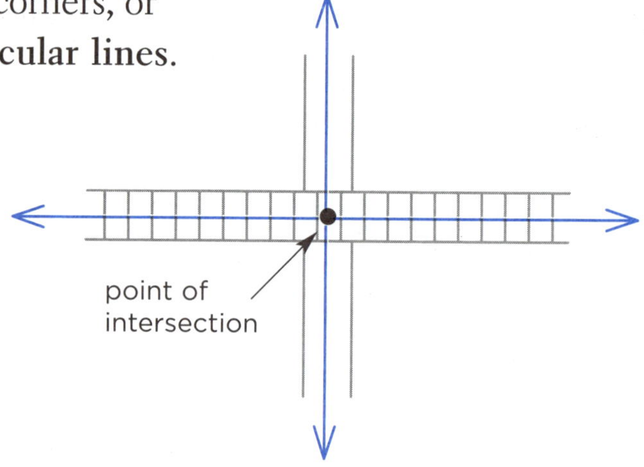

1. Are perpendicular lines also intersecting lines? Explain.

____

____

2. Line segments can also intersect. These line segments intersect at point *P*. Do you think these line segments are perpendicular? Explain.

____

____

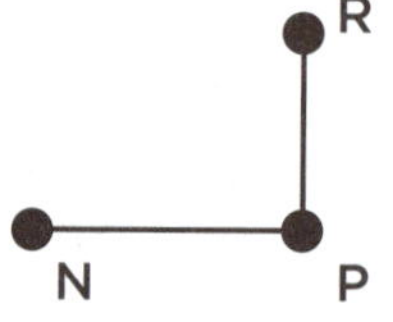

Write *intersecting* or *perpendicular* to best describe each pair of lines below.

3.

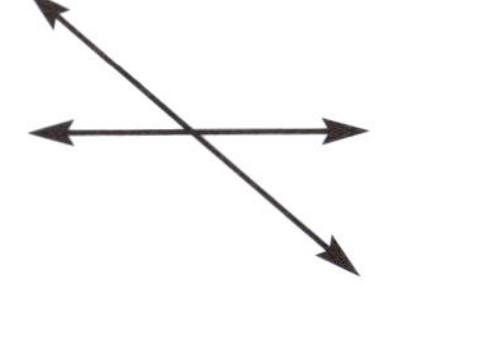

____

4.

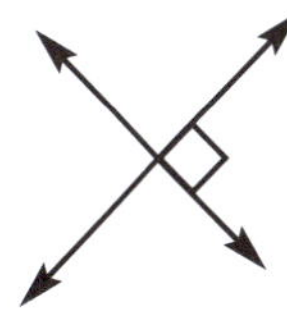

____

5.

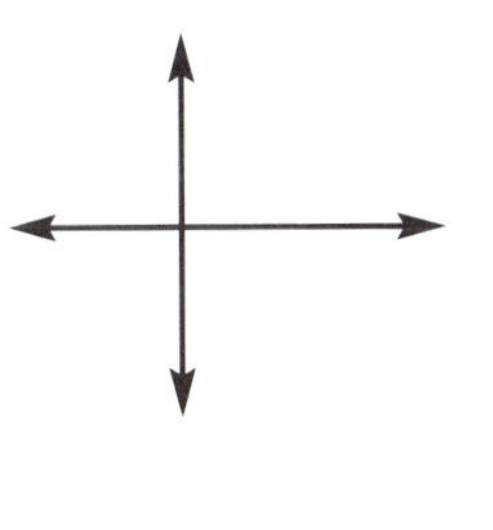

____

6.

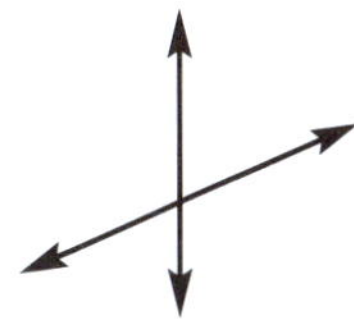

____

7. In the space at the right, draw a pair of intersecting lines that are **not** perpendicular.

Solve a Problem

8. Find and list three objects in your classroom that represent perpendicular lines.

____

____

It's a Fact!

About 80% of the Central Pacific Railroad workers were Chinese immigrants.

Parallel Lines

The two rails of a railroad track must always be the same distance apart. This keeps a train's wheels moving along the track. Railroad tracks can represent parallel lines.

Parallel lines are always the same distance apart. They never cross, or intersect, one another. The line segments that make up parallel lines are also parallel.

Follow the steps below to draw parallel lines.

STEP 1 Use a ruler to draw a line. Remember to add arrows to the ends to show that it is a line.

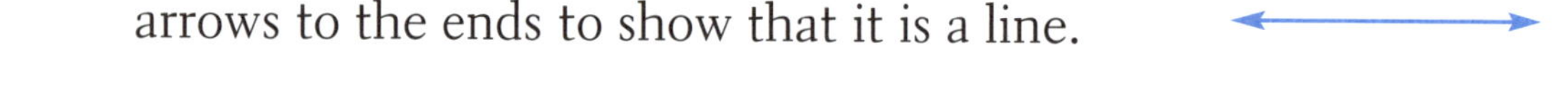

STEP 2 Draw a point one inch below the line. Use a ruler to measure.

STEP 3 Draw another point one inch below the line.

STEP 4 To draw a parallel line, use your ruler to connect the points you drew. Add arrows to the ends to show that it is a line.

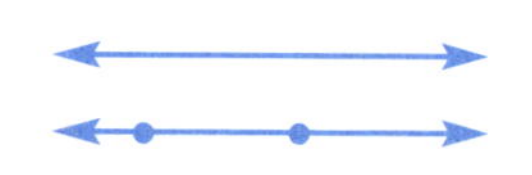

Use your ruler to complete the following tasks.

1. Draw a line parallel to this line. Draw the lines $\frac{1}{2}$ inch apart.

____

2. Draw a line parallel to this line. Draw the lines $\frac{1}{4}$ inch apart.

____

3. Draw a line that is **not** parallel to this line.

4. Tell why the lines in Problem 3 are **not** parallel lines.

____

____

____

____

On Your Own

Look for three objects in your neighborhood that represent parallel lines. Describe them.

1. Which of these figures contains one pair of perpendicular line segments?

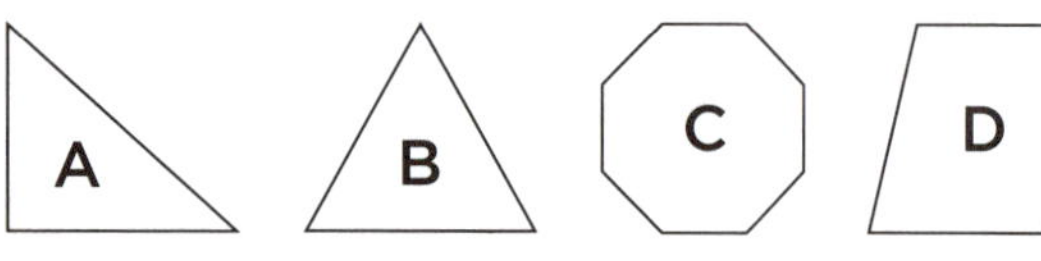

Ⓐ Figure A   Ⓒ Figure C

Ⓑ Figure B   Ⓓ Figure D

2. What kind of lines do railroad tracks represent?

Ⓕ parallel lines

Ⓖ perpendicular lines

Ⓗ intersecting lines

Ⓙ They don't represent any kind of lines.

3. What kind of lines are shown below?

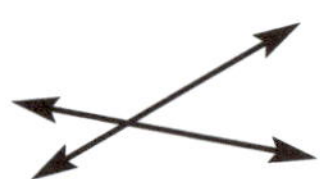

Ⓐ parallel lines

Ⓑ perpendicular lines

Ⓒ intersecting lines

Ⓓ They don't represent any kind of lines.

4. What kind of angles form when perpendicular lines intersect?

____

5. Draw a pair of parallel lines.

Write two statements to describe the lines you drew.

____

____

____

6. Think Back Write *acute*, *obtuse*, or *right* to describe the angle shown below.

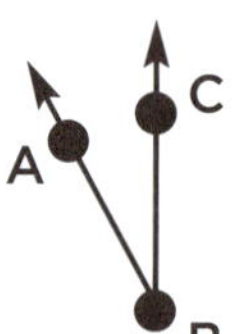

____

Flag Day

In 1949, President Truman signed an act declaring June 14 as Flag Day. That's because on June 14, 1777, the Continental Congress passed the first Flag Act. The act called for a flag with 13 red and white stripes and 13 white stars on a blue background. Over the years, the flag has kept its stripes, but added one new star for each state. The last star, representing Hawaii, was added on July 4, 1960.

Get Started

For a social studies project, Mr. West's class studied family histories. Some students drew flags to show the countries their relatives had come from.

Anna drew a flag with three sections. She colored the bottom section red and the top left section blue. She left the star and the top right section white. Use crayons to color Anna's flag.

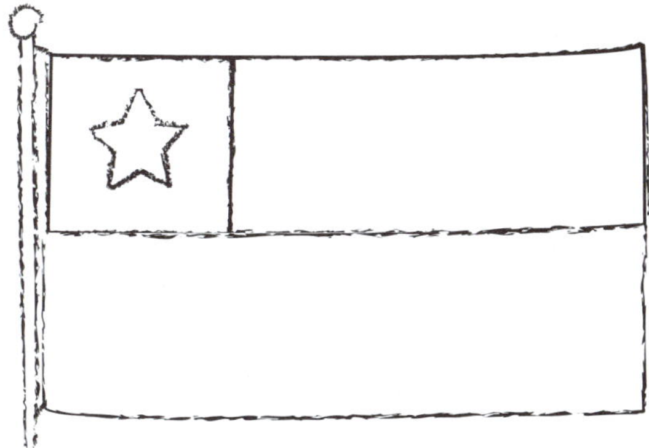

- What shape does the flag represent? ___________________

- What shapes do you see inside the flag?

____

- What country does this flag represent? Use a reference book to find the answer. ___________________

Quadrilaterals

A polygon with four sides is called a **quadrilateral**. A quadrilateral also has four angles.

Not all quadrilaterals look alike. And some quadrilaterals can have more than one name. Study the diagram below carefully to learn about some special kinds of quadrilaterals.

Parallelogram

For a quadrilateral to be a **parallelogram,** it must have opposite sides that are parallel.	

Rectangle

This polygon has opposite sides that are parallel, so it is also a parallelogram. But it has 4 right angles, which makes it a **rectangle**.

Rhombus

A **rhombus** is also a parallelogram, because its opposite sides are parallel. What makes it a rhombus is that its sides are all equal.

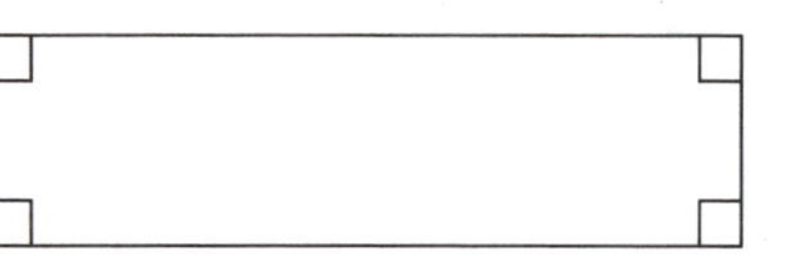

Square

A **square** has opposite sides that are parallel, 4 right angles, and 4 sides that are all the same length. So a square is a parallelogram, a rectangle, and a rhombus!

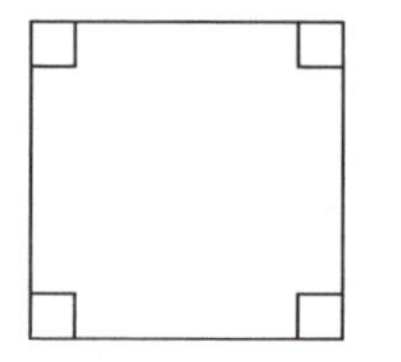

1. The figure at the right is a **trapezoid.** Why is this figure **not** a parallelogram?

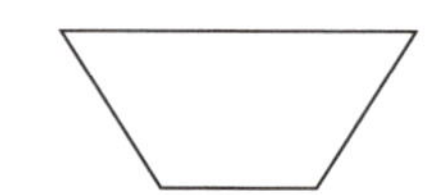

____

____

Write all the names for each quadrilateral shown.

2.

____

____

____

3.

____

____

____

4.

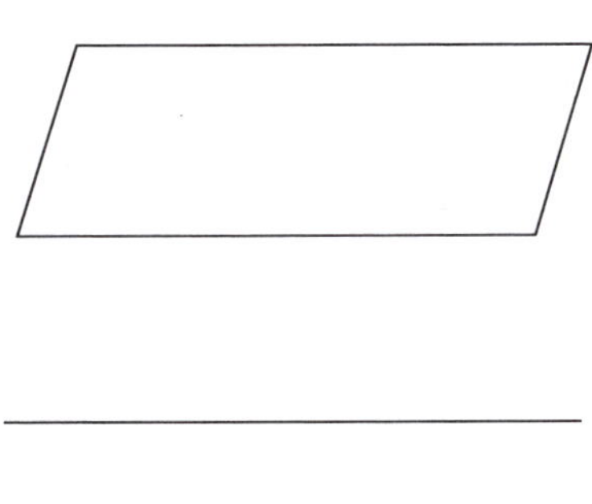

____

____

____

5.

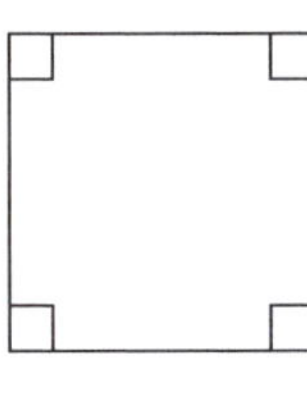

____

____

____

6. These two parallelograms each have 4 right angles:

_________________ and _________________

7. These two parallelograms each have 4 equal sides:

_________________ and _________________

8. Explain why a circle is **not** a polygon.

____

____

It's a Fact!

Francis Hopkinson, a Congressman from New Jersey, is believed to have designed the first Stars and Stripes.

Other Polygons

Polygons with different numbers of sides have different names. Look at the different kinds of polygons in Gilbert's flag. It has one polygon with 3 sides, five polygons with 4 sides, and one polygon with 10 sides.

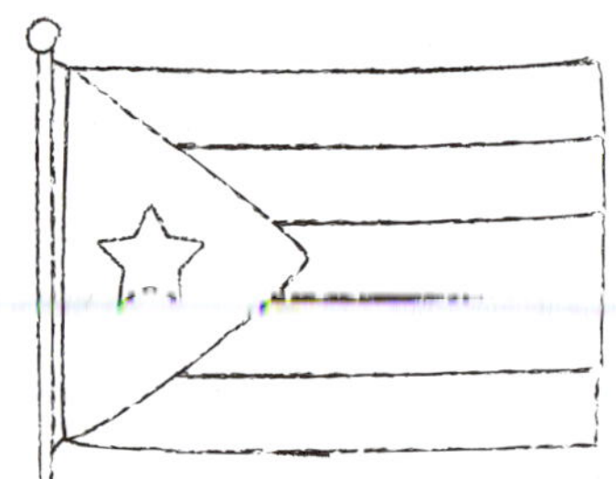

The first few letters, or prefix, of a polygon's name can tell you how many sides it has. Look at these examples.

Prefix	Number of Sides	Polygon
Tri-	3	Triangle
Quadri-	4	Quadrilateral
Penta-	5	Pentagon
Hexa-	6	Hexagon
Octa-	8	Octagon
Deca-	10	Decagon

Use the table above and follow the steps below to draw an octagon.

STEP 1 Check the table to see how many sides an octagon has.

STEP 2 Remember, a polygon is a closed figure made of line segments. How many line segments, or sides, does an octagon have? __________

STEP 3 Draw a picture of your octagon.

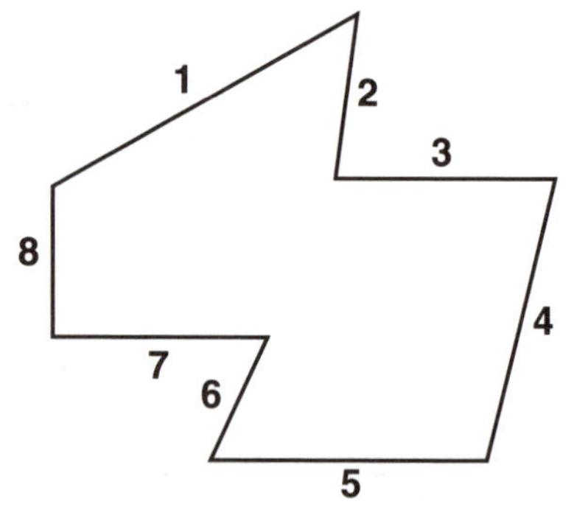

Follow the steps you learned to draw an example of
each polygon.

1. hexagon

2. triangle

3. pentagon

4. decagon

Count the sides of each of these polygons.
Write their names.

5. 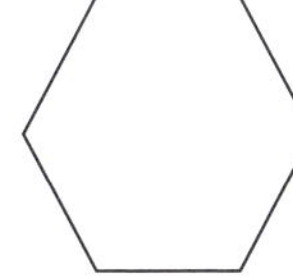 _______________________

6. 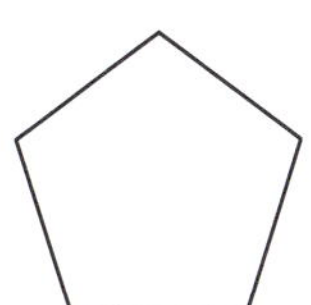 _______________________

On Your Own

Use what you have learned
about polygons to design a
polygon flag for your class.
Draw a picture of your design.

10 Test Yourself

1. How many sides does a
quadrilateral have?

Ⓐ 8   Ⓑ 6   Ⓒ 4   Ⓓ 3

2. Which of these polygons is a
quadrilateral?

Ⓕ parallelogram

Ⓖ pentagon

Ⓗ hexagon

Ⓙ triangle

3. Which of these quadrilaterals has
sides that are all the same length?

Ⓐ parallelogram

Ⓑ rectangle

Ⓒ rhombus

Ⓓ trapezoid

4. Write the names of two
quadrilaterals that each
have 4 right angles?

____

____

5. Name 3 polygons that each have
at least one pair of parallel sides.

____

____

____

6. Think Back Write this fraction
as a decimal to the hundredths
place. Show your work.

$\dfrac{3}{5} =$

Ones	.	Tenths	Hundredths
0	.		

64   Level D

Inventing Measurement

How did people build pyramids for the pharaohs, or kings, of ancient Egypt? They used tools, of course, but not the same tools we use today. They used a tool called a cubit. It was the same length as the pharaoh's forearm, plus the width of his palm. But with each new pharaoh the length of the cubit had to change. Today, we use standard units of measure that don't change. This makes it easier to build anything—even a pyramid.

Get Started

Students in Ms. Sanchez's class were using their centimeter rulers to measure objects around the classroom. Below are some of the objects they measured.

First, estimate the length of the object. Then use your centimeter ruler to measure. Find the centimeter mark on the ruler closest to the end of the object.

> **Remember**
> The width of your pinky finger is about 1 centimeter.

Estimate

_______ cm

Measure

_______ cm

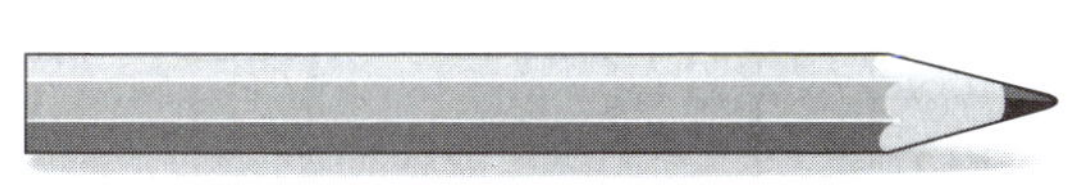

Estimate

_______ cm

Measure

_______ cm

Estimate

_______ cm

Measure

_______ cm

Working with Perimeter

The measurement of the distance around the outside of a figure is called the perimeter of the figure.

Look at the figures shown on the grids below. You can find the perimeter of each figure by counting the number of units around the outside of each figure.

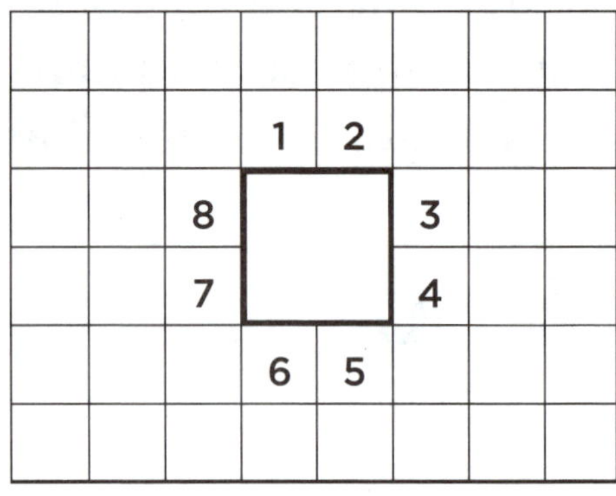

The perimeter of this square is 8 units.

Each unit has the same length. You can imagine that the units are inches, meters, or even miles.

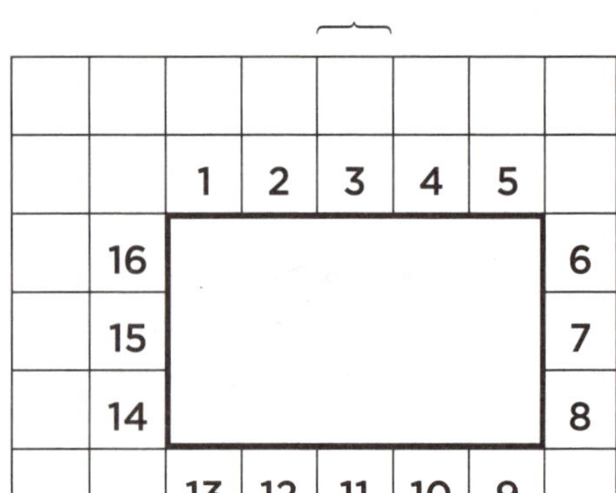

The perimeter of this rectangle is 16 units.

---

1. Look at the square shown on the grid at the right. Find the perimeter of this square.

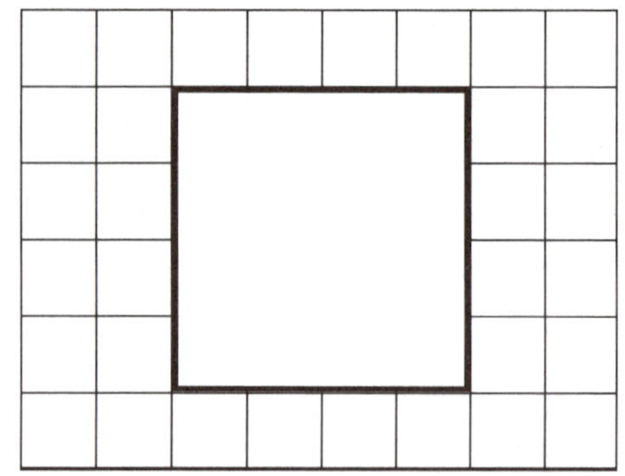

Perimeter = _______ units

Find the perimeter of each figure below.

2.

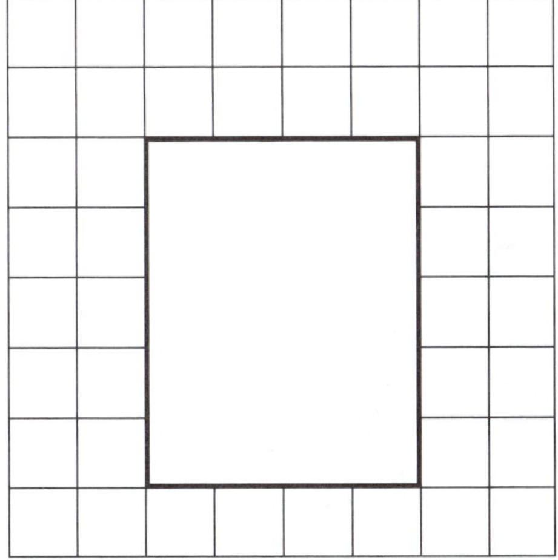

Perimeter = ________ units

3.

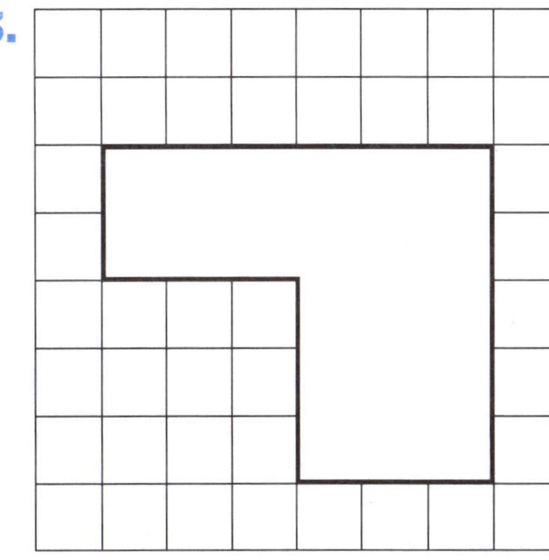

Perimeter = ________ units

4.

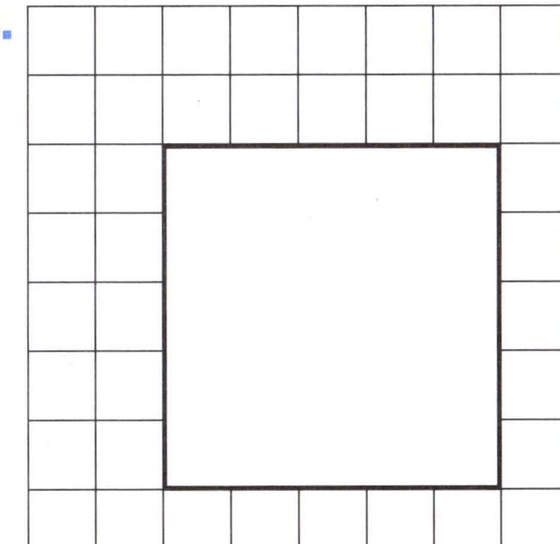

Perimeter = ________ units

5.

Perimeter = ________ units

Solve a Problem

6. Draw a figure with a perimeter of 20 units on the grid below.

It's a Fact!

The prefix *peri* means "around," and *meter* means "measure."

So *perimeter* means "the measure around" a figure.

Measuring Perimeter

Ms. Sanchez's class had their picture taken. They wanted
to decorate the picture by putting ribbon around the
outside edge.

Follow the steps below to find the perimeter of the picture,
so you will know the length of ribbon that is needed.

STEP 1 Use your centimeter ruler to
measure each side. Record
each measure.

_______ cm

_______ cm   _______ cm

_______ cm

STEP 2 Add the measures of all the sides.

_______ cm + _______ cm + _______ cm + _______ cm = _______ cm

STEP 3 Write the perimeter.

Perimeter = _______ cm

Follow the steps you learned to find the perimeter
of each figure below.

1. 

Perimeter = _________ cm

2. 

Perimeter = _________ cm

3. 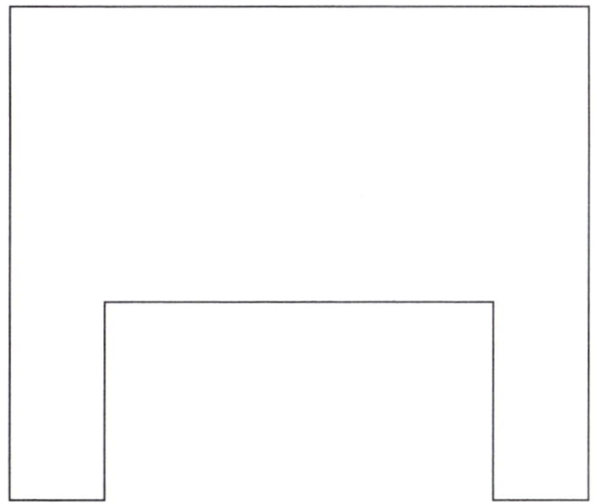

Perimeter = _________ cm

4. 

Perimeter = _________ cm

5. How can you find the perimeter of a
rectangle by only measuring two sides?

____

____

____

____

On Your Own

A square has a perimeter
of 32 cm. What is the
measure of each side?

1. What is the perimeter of the square garden shown?

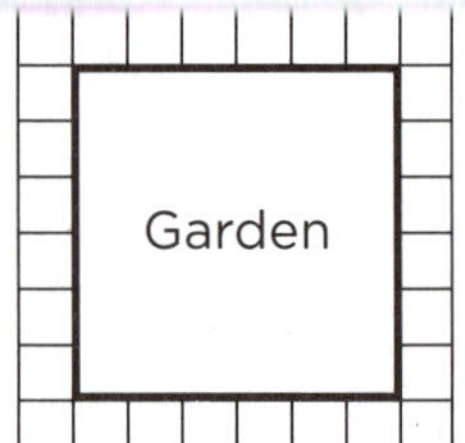

Ⓐ 12 units   Ⓒ 24 units

Ⓑ 18 units   Ⓓ 36 units

2. What is the perimeter of the rectangle?

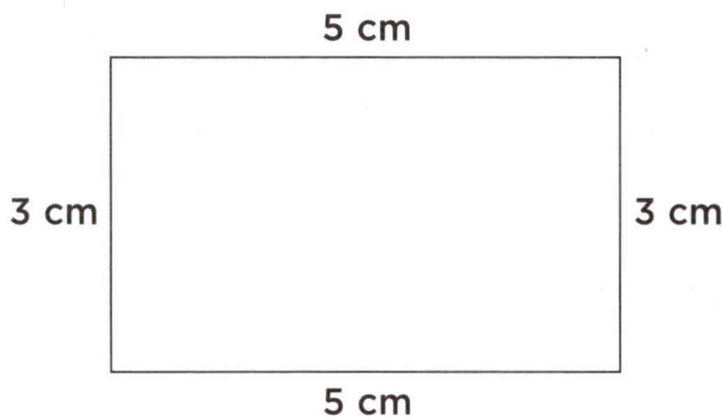

Ⓕ 8 cm   Ⓗ 30 cm

Ⓖ 16 cm   Ⓙ 34 cm

3. What is the perimeter of the triangle shown below?

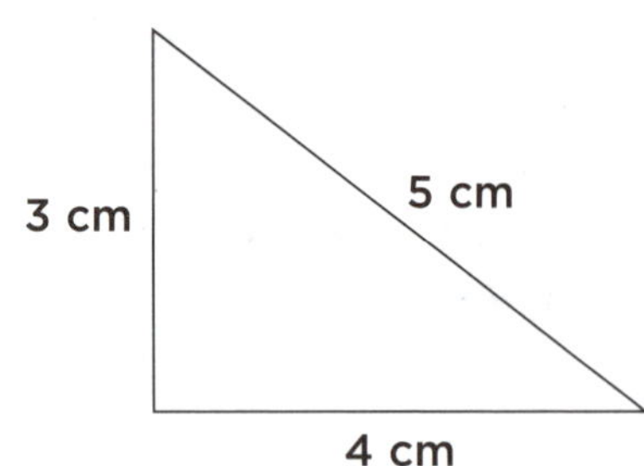

Ⓐ 9 cm   Ⓒ 17 cm

Ⓑ 12 cm   Ⓓ 60 cm

4. Which figure does **not** have a perimeter of 36 centimeters?

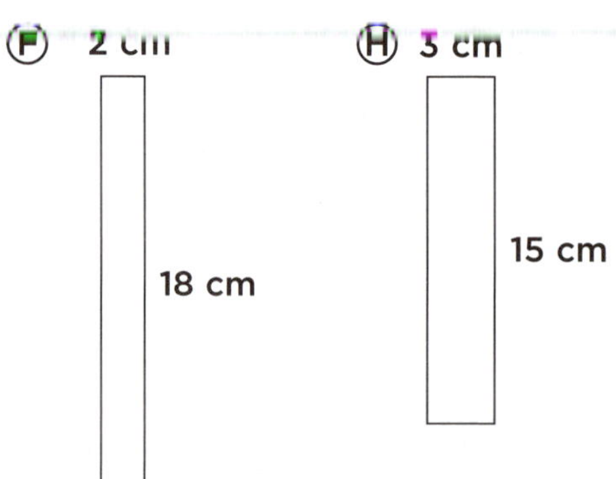

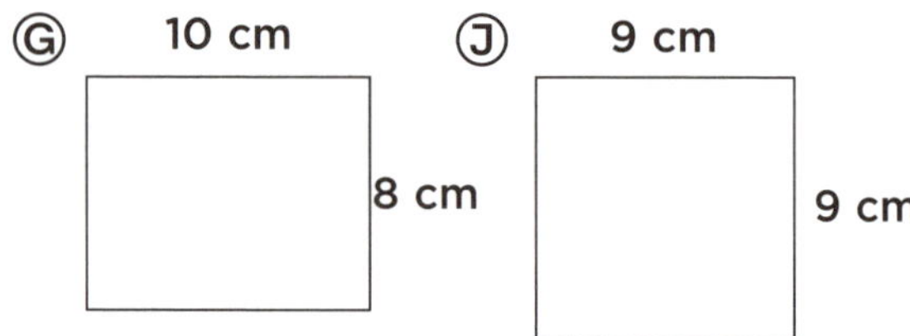

5. On the lines below, explain how you found your answer to Question 4 above.

____

____

____

____

6. Think Back Name 3 shapes that each contain at least one pair of parallel line segments.

____

____

____

Postage Stamp Quilt

Imagine cutting 69,649 tiny squares of fabric, each the size of a postage stamp. Then imagine using more than 2 miles of thread to stitch these tiny squares together to make a quilt. That's what Mrs. B. W. Riley did in 1939. It took her 425 hours to cut the fabric. Then it took another 1,087 hours to sew the pieces together. Mrs. Riley's postage stamp quilt won first prize in the 1940 Iowa State Fair.

Get Started

Brent and Candace are making a postage stamp quilt. It is not finished yet, but the drawing below shows how big the quilt will be when it is finished.

Draw lines to show what the completed quilt will look like.

- How many square patches are in each row? ________

- How many square patches are in each column? ________

- How many square patches will there be in all? ________

Working with Area

Look at the rectangle at the right.
Each small square is a square unit. ⟶

There are 12 square units that
cover the rectangle. They form
3 rows and 4 columns.

1	2	3	4
5	6	7	8
9	10	11	12

Area = 12 square units

The **area** of a figure is the number of square units
that cover the figure without overlapping. So the
area of the rectangle above is 12 square units.

Look at the figure at the right.

There are 16 square units that
cover the figure. They form 4 rows
and 4 columns.

The area of the figure is 16 square units.

1	2	3	4
5	6	7	8
9	10	11	12
13	14	15	16

Area = 16 square units

---

1. Look at the rectangle below.

Think:
You can imagine that the
square units are square
centimeters, square feet,
or even square miles.

There are _________ square units that cover the rectangle.

They form _________ rows and _________ columns.

The area of the rectangle is _________ square units.

Tell how many rows and columns are formed by the square units covering each figure below. Then find the area of each figure.

2.

_____ rows and _____ columns

Area = _____ square units

3.

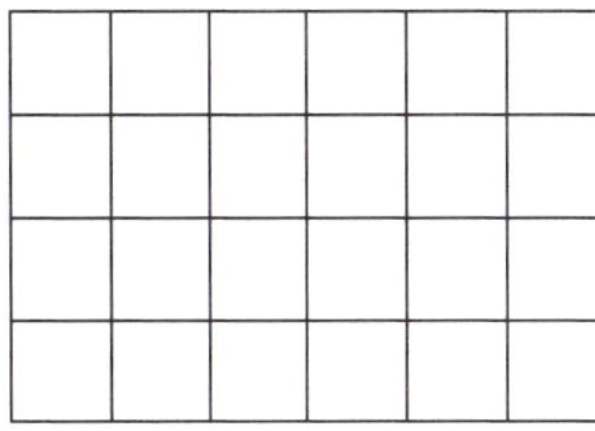

_____ rows and _____ columns

Area = _____ square units

4.

_____ rows and _____ columns

Area = _____ square units

5.

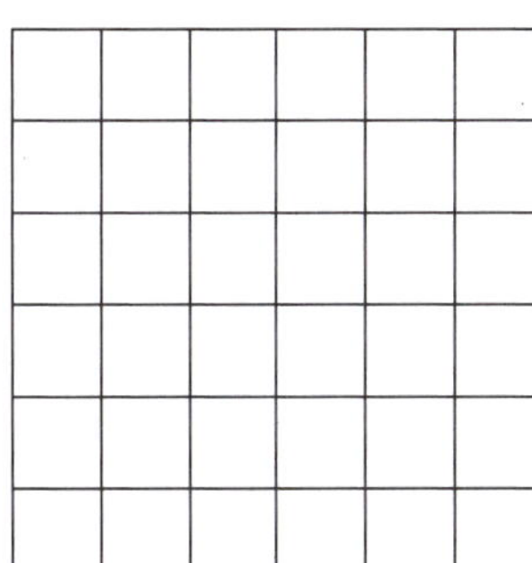

_____ rows and _____ columns

Area = _____ square units

6. On the grid at the right, draw a figure with an area of 18 square units.

Multiply to Find Area

At a class trip to the museum, Mr. Tang's class saw an exhibit of beautiful patchwork quilts. One of the quilts they saw was 6 feet long and 4 feet wide.

How can you find the area of this quilt?

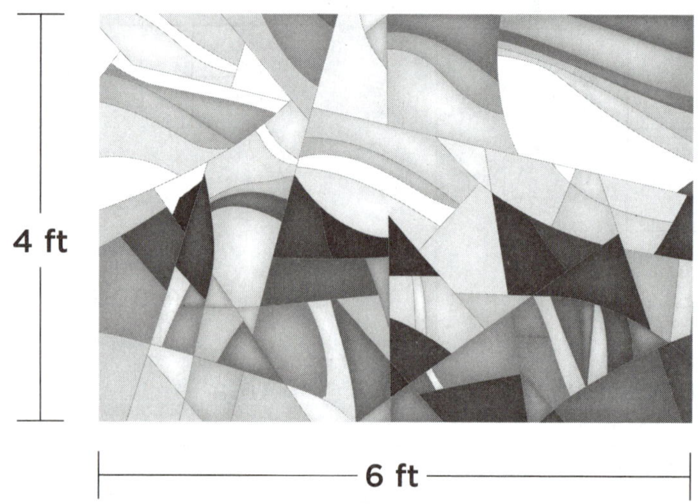

There is a way to find the area of a rectangle without covering it with square units and counting them.

Follow the steps below.

STEP 1 Record the length and the width of the rectangle.

length = ________ ft   width = ________ ft

STEP 2 Multiply the length and the width.

________ ft × ________ ft = ________ square feet
   length   width   area

The product is the number of square units that cover the rectangle. Since the units we are using are feet, the area is expressed in square feet.

Use the steps you learned to find the area of each figure below.

1.

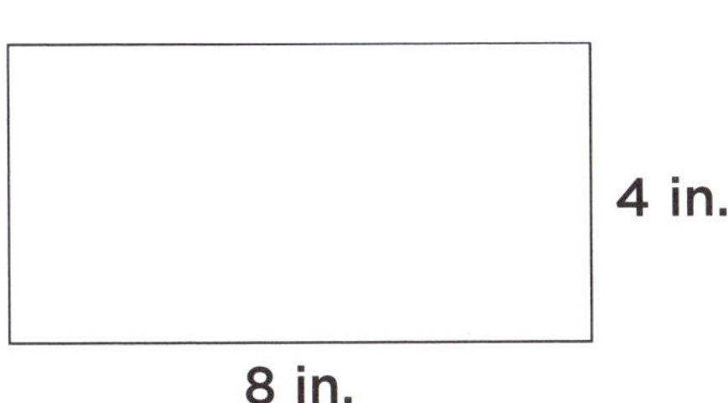

Area = _____ square inches

2.

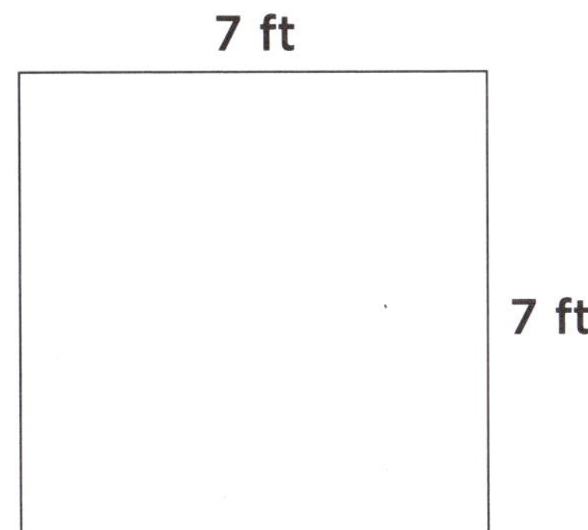

Area = _____ square feet

3.

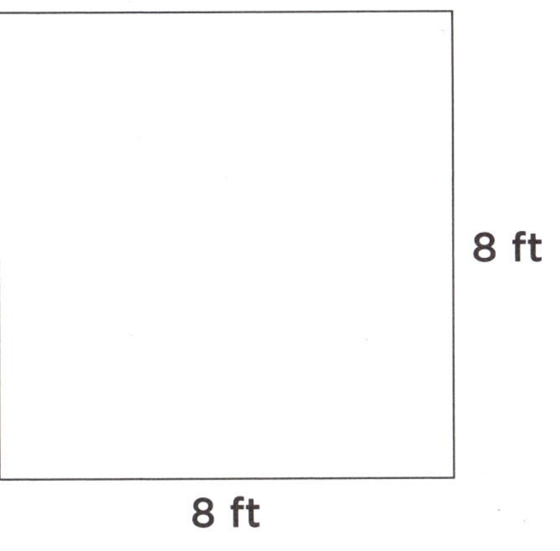

Area = _____ square feet

4.

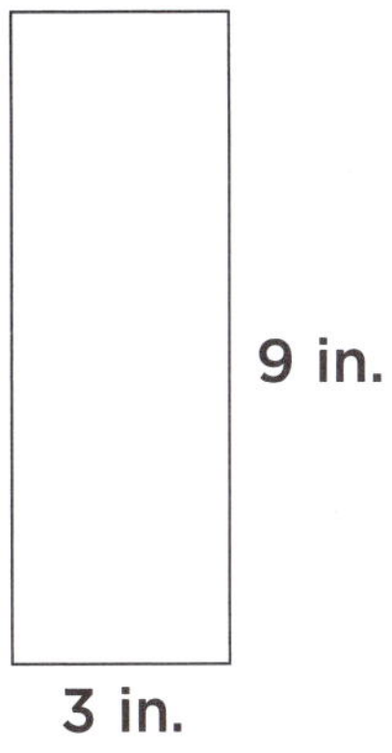

Area = _____ square inches

It's a Fact!

Patchwork quilts are often made from leftover scraps of fabric. Some people use scraps of worn-out clothing, including military uniforms.

On Your Own

Suppose a rectangular quilt is 8 feet long and has an area of 40 square feet.

What is the width of the quilt?

1. What is the area of the rectangle shown below?

   Ⓐ 11 square units

   Ⓑ 22 square units

   Ⓒ 28 square units

   Ⓓ 56 square units

2. Which can be used to find the area of a rectangle?

   Ⓕ length + width

   Ⓖ length × width

   Ⓗ length − width

   Ⓙ length ÷ width

3. What is the area of the rectangle shown below?

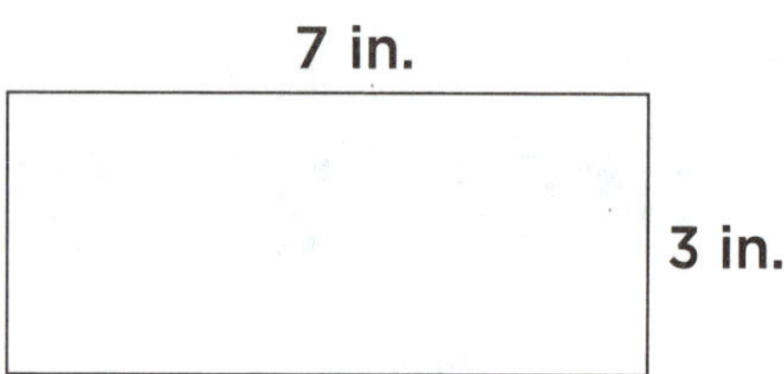

   Ⓐ 30 square inches

   Ⓑ 21 square inches

   Ⓒ 20 square inches

   Ⓓ 10 square inches

4. A rectangle measures 8 yards by 9 yards. What is its area? Show your work below.

   Area = _______________

5. On the grid below, draw a rectangle with an area of 32 square units.

6. Think Back Find the perimeter of the figure you drew above.

   Perimeter = _______________

The Inchworm

Some moths that flutter around lights at night have broad wings marked with thin, wavy lines. But they didn't always look like this. As caterpillars, they may have been called inchworms, measuring worms, or loopers. That's because of the way these caterpillars move. They bring their back to the front, making a loop in the middle. Then the back holds on while the front moves forward, stretching the length of their body—about one inch.

Get Started

Estimate the length of each object below. Then measure with your inch ruler. Find the inch mark closest to the end of the object.

Remember
A quarter is about 1 inch wide.

Estimate   **Measure**

_______ in.   _______ in.

Estimate   **Measure**

_______ in.   _______ in.

Estimate   **Measure**

_______ in.   _______ in.

Working with Half Inch and Quarter Inch

When you measure an object, sometimes the end of the object is between two inch marks. You can measure to the nearest inch, as you did on page 77. You can also measure to the nearest half inch or quarter inch.

Look at the examples below.

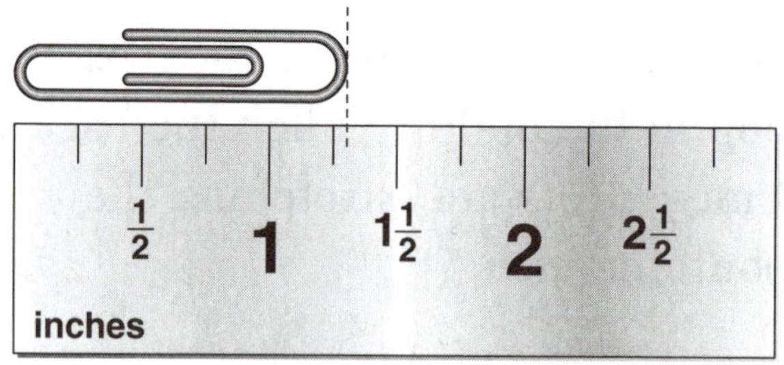

On this ruler, all of the half-inch marks are labeled.

The end of the paper clip is between two half-inch marks—1 in. and $1\frac{1}{2}$ in. It is closer to $1\frac{1}{2}$ in., so the length to the nearest half inch is $1\frac{1}{2}$ in.

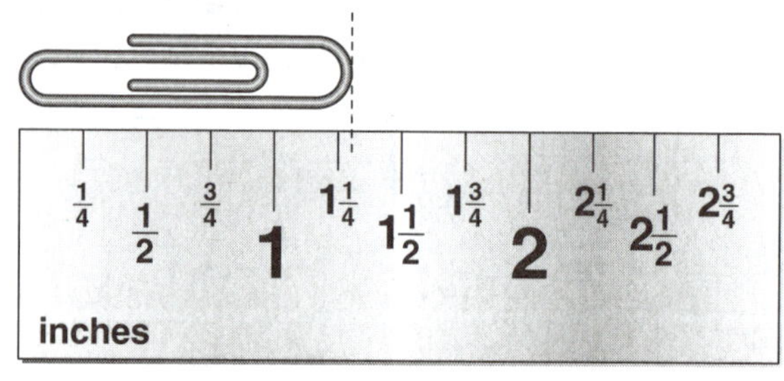

On this ruler, all of the quarter-inch marks are labeled.

The end of the paper clip is between two quarter-inch marks—$1\frac{1}{4}$ in. and $1\frac{1}{2}$ in. It is closer to $1\frac{1}{4}$ in., so the length to the nearest quarter inch is $1\frac{1}{4}$ in.

1. The same paper clip is used in each example above.

To the nearest half inch, the paper clip measures _________ in.

To the nearest quarter inch, the paper clip measures _________ in.

Use what you learned to find the measure of each object below to the nearest half inch and to the nearest quarter inch.

2.

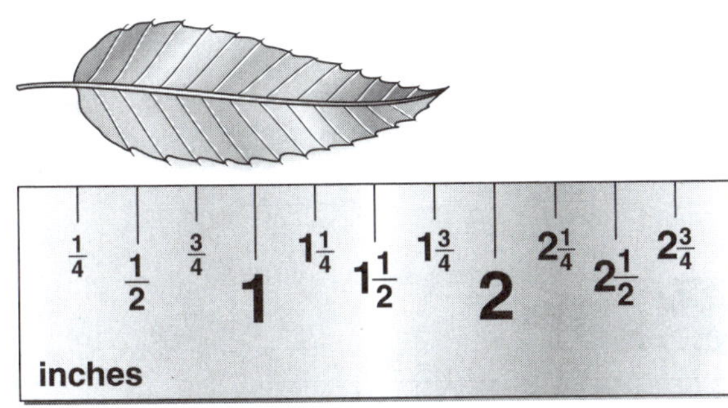

The length of the leaf to the nearest . . .

half inch is ___________ in.

quarter inch is ___________ in.

3.

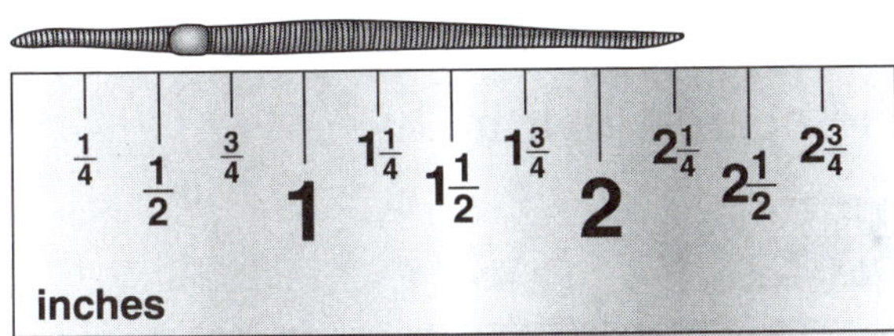

The length of the worm to the nearest . . .

half inch is ___________ in.

quarter inch is ___________ in.

4. Linda found an earthworm in her backyard. When she measured it, she found that the worm was $3\frac{3}{4}$ inches long. Use your ruler to draw a picture of the worm below.

It's a Fact!

One of the largest earthworms in the world is the Giant Gippsland Earthworm of Australia.

These worms can be over 3 feet long!

Measure with Half Inch and Quarter Inch

Look at the crayon below. On page 77, you measured its length to the nearest inch. What is the length to the nearest half inch and to the nearest quarter inch? You can follow the steps below to find out.

STEP 1 Start at the "0" end of your ruler. Mark the end of the object above the ruler.

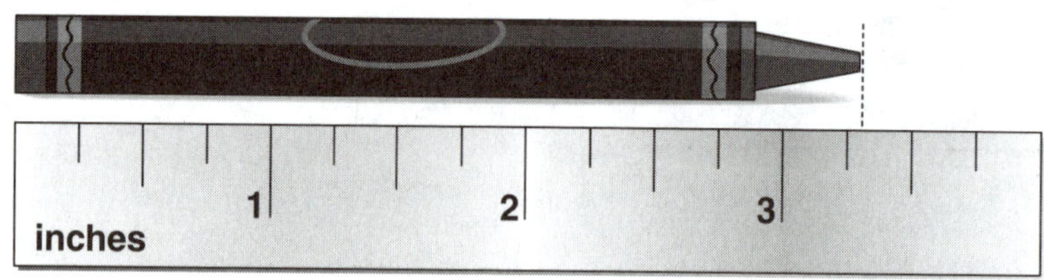

STEP 2 Find the half-inch mark closest to the end of the object. This is the length to the nearest half inch.

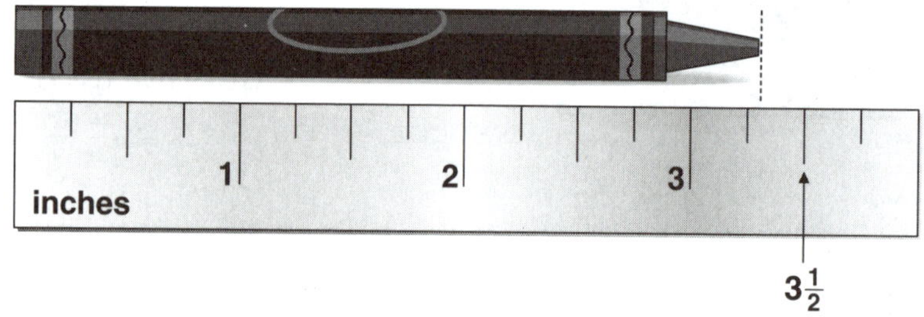

STEP 3 Find the quarter-inch mark closest to the end of the object. This is the length to the nearest quarter inch.

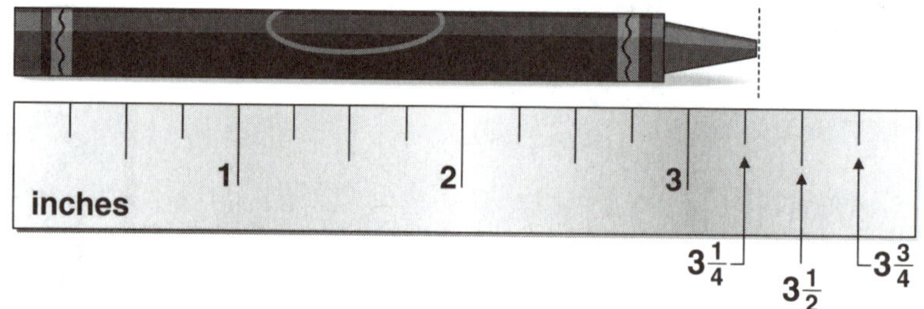

What is the length of the crayon to the nearest

half inch? _____________ quarter inch? _____________

Use your ruler and the steps you learned to measure the lengths of the objects below to the nearest half inch or quarter inch.

1.

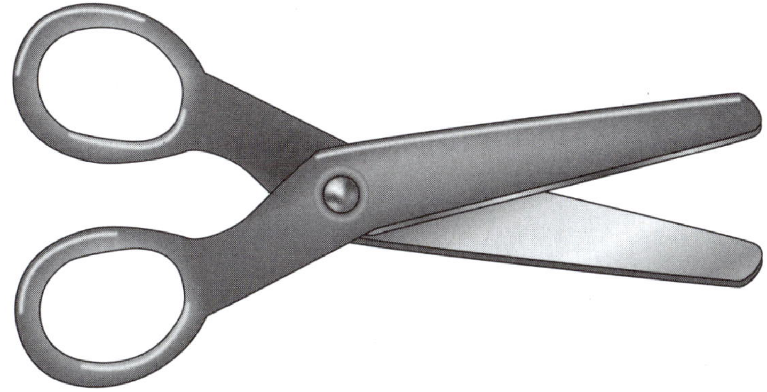

The length to the nearest quarter inch is ______________ .

2.

The length to the nearest half inch is ______________ .

3.

The length to the nearest quarter inch is ______________ .

4. Label these marks on the ruler:

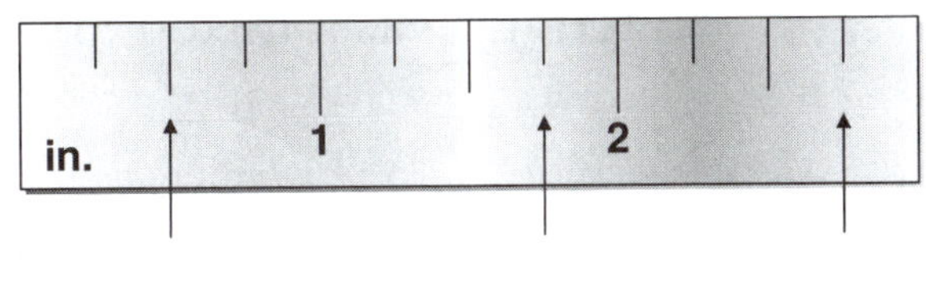

______________   ______________   ______________

On Your Own

On a sheet of paper, draw lines with these lengths:

$2\frac{1}{2}$ in. $1\frac{3}{4}$ in.

$\frac{1}{2}$ in. $4\frac{1}{4}$ in.

13 Test Yourself

1. What is the length of the eraser to the nearest quarter inch?

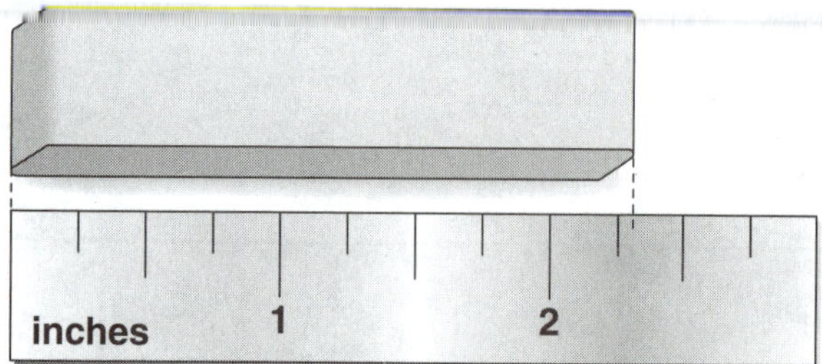

Ⓐ $1\frac{1}{2}$ in.

Ⓑ 2 in.

Ⓒ $2\frac{1}{4}$ in.

Ⓓ $2\frac{1}{2}$ in.

2. What is the length of the leaf to the nearest half inch?

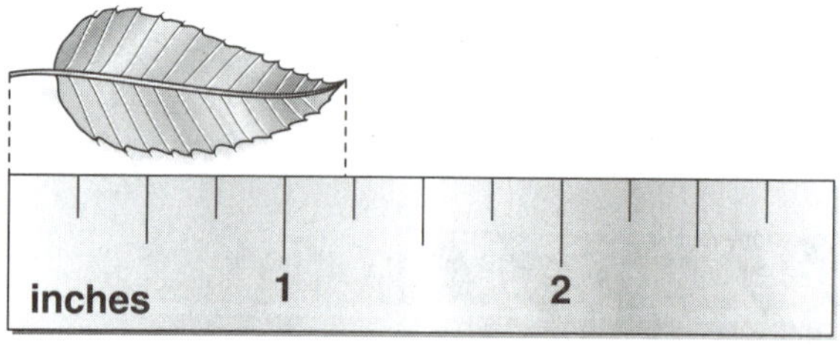

Ⓕ $\frac{1}{2}$ in.

Ⓖ 1 in.

Ⓗ $1\frac{1}{4}$ in.

Ⓙ $1\frac{1}{2}$ in.

3. Use your ruler. Which line is closest to $2\frac{1}{2}$ in. long?

Ⓕ _________________

Ⓖ ___________________

Ⓗ _____________________

Ⓙ _______________________

4. Measure the nail below to the nearest half inch and quarter inch.

nearest half inch: _________

nearest quarter inch: _________

5. Use your inch ruler to draw a line segment that measures $2\frac{1}{4}$ inches long.

6. Think Back Write the improper fraction $\frac{15}{4}$ as a mixed number.

$\frac{15}{4} =$ _____________

Shaquille O'Neal

Shaquille O'Neal, also known as Shaq, is a famous basketball player. Although most basketball players are very tall, Shaq is among the tallest at about 7 feet 1 inch in height. He weighs about 340 pounds. In case you wonder what size shoe such a tall person would wear, Shaq wears size 21EEE!

Get Started

Follow the steps below to find out how tall you are. You will need a ruler and a partner.

STEP 1 Stand with your back against the chalkboard or the wall. Have your partner make a mark to show your height.

STEP 2 Use your ruler to find out how tall you are in feet and inches. Write your name and height in the chart below.

STEP 3 Have your partner follow the steps above to find his or her height.

Name	Height
	_____ feet _____ inches
	_____ feet _____ inches

Working with Equivalent Measurements

You can use different units of measure to describe how long or tall something is.

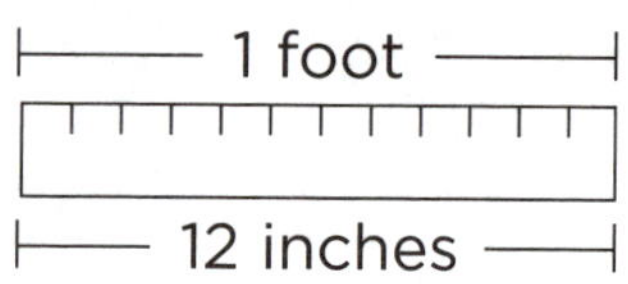

Look at the picture at the right. It represents a ruler that is 1 foot long. You can also say that it is 12 inches long.

If you put two of these rulers end to end, they would measure 2 feet in all. Two feet equals 24 inches, because there are two groups of 12 inches.

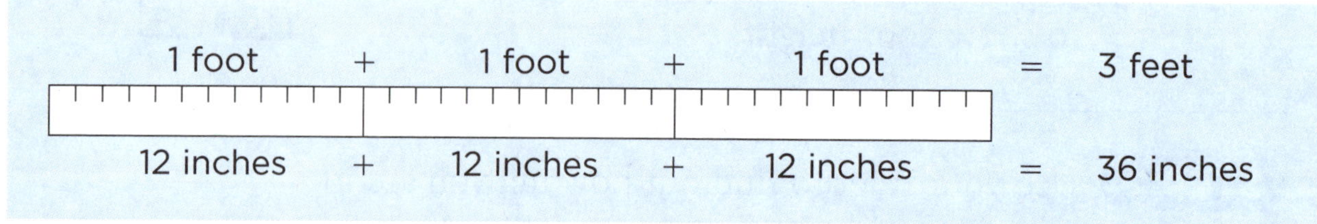

If you add a third ruler, the three rulers end to end would measure 3 feet, or 36 inches, in all.

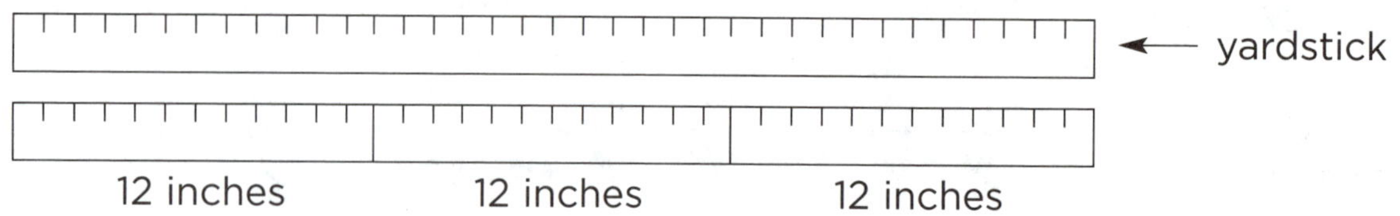

1. Look at this model of a yardstick.

How many feet are in 1 yard? _________ feet

How many inches are in 1 yard? _________ inches

Rename each measurement below.

2.

How many inches
long is the puppy? _____ inches

3.

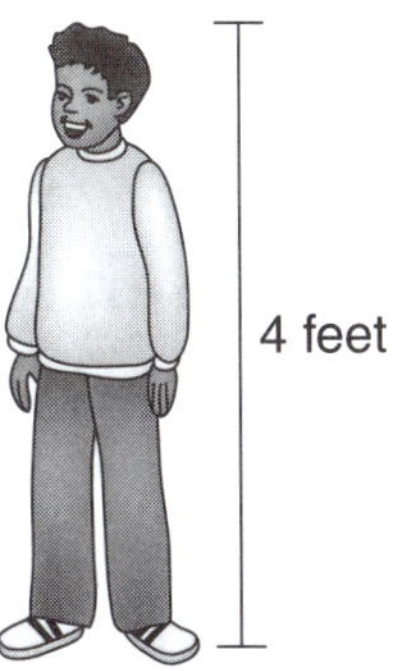

How many inches
tall is the student? _____ inches

4.

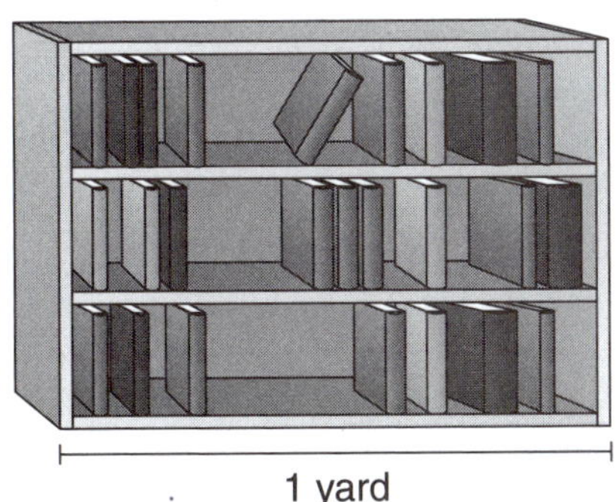

How many feet wide
is this bookcase? _____ feet

5.

How many inches
deep is this pool? _____ inches

6. Jenna is 4 feet 3 inches tall.
How tall is Jenna in inches?
Show your work below.

It's a Fact!

Shaquille O'Neal was born in Newark, New Jersey, on March 6, 1972.

Jenna is _____ inches tall.

Finding Equivalent Measurements

Shaquille O'Neal is about 7 feet 1 inch tall. What is this height in inches? The rim of the basketball hoop is 3 yards 1 foot from the floor. What is this height in feet?

Follow the steps below to rename these measurements.

Feet to Inches	**Yards to Feet**
STEP 1 Record the measure. 7 ft 1 in.	**STEP 1** Record the measure. 3 yd 1 ft
STEP 2 Multiply the number of feet by 12 inches. 7×12 in. = ______ in.	**STEP 2** Multiply the number of yards by 3 feet. 3×3 ft = ______ ft
STEP 3 Add to find the total height. 84 in. + 1 in. = 85 in.	**STEP 3** Add to find the total height. 9 ft + 1 ft = 10 ft

Why can you use multiplication to change feet to inches and yards to feet?

____

____

Use the steps you learned to rename the
measurements below.

1. Juan's father is 6 feet tall.
How many inches tall is he?

______ inches

2. A pool is 12 yards long. How
many feet long is the pool?

______ feet

3. Theresa jumped a distance
of 5 feet. How many inches
did she jump?

______ inches

4. The see-saw area at the
park is 5 yards 2 feet wide.
How many feet wide is this?

______ feet

5. Marcus is 4 feet 6 inches tall.
How many inches tall is he?

______ inches

6. Sonia's tree is 3 feet 8 inches tall.
How many inches tall is it?

______ inches

On Your Own

The hallway in Olga's
house is 2 yards 2 feet
2 inches long.

How many inches long
is the hallway?

1. How many inches are there in 1 yard?

Ⓐ 3

Ⓑ 12

Ⓒ 18

Ⓓ 36

2. Which measurement below is the longest?

Ⓕ 17 inches

Ⓖ 1 foot 7 inches

Ⓗ 2 feet

Ⓙ 1 foot 6 inches

3. Maria is 4 feet 8 inches tall. How many inches tall is she?

Ⓐ 32 inches

Ⓑ 48 inches

Ⓒ 52 inches

Ⓓ 56 inches

4. Which of these measurements is the shortest: 5 feet, 2 yards, or 73 inches? Show your work.

____

5. Randy's bedroom is 6 yards long. How many feet long is this? Show your work.

_________ feet long

6. **Think Back** Find the product. Show your work below.

$$27 \times 8 = \underline{\hspace{2cm}}$$

Zulu Beads

Imagine using beads to send a message. The Zulu people of southern Africa do. They combine geometric shapes and colorful beads to express facts and feelings. Black beads can say someone is sad. Blue beads can ask a question. Red and white beads can stand for love. Yellow, pink, and green beads have meaning, too. Each pattern is a code for others to read. By using repeating patterns of beads and shapes, the Zulu people can speak without saying a word.

Get Started

After completing a social studies unit on Africa, Mr. Arroyo's class decided to design and make jewelry. Each group of students used a different repeating pattern to create a beaded bracelet.

For each bracelet below, draw the next three beads in the pattern.

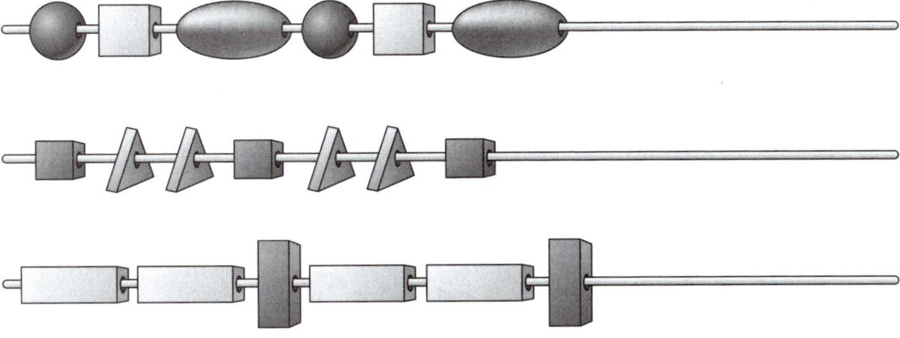

How did you know how to continue the pattern in each bracelet?

____

____

____

Working with Patterns

Some patterns do not repeat. Instead, they may increase (grow) or decrease (shrink) in a predictable way. To predict what comes next in a pattern, you must understand how that pattern works.

Look at this pattern.

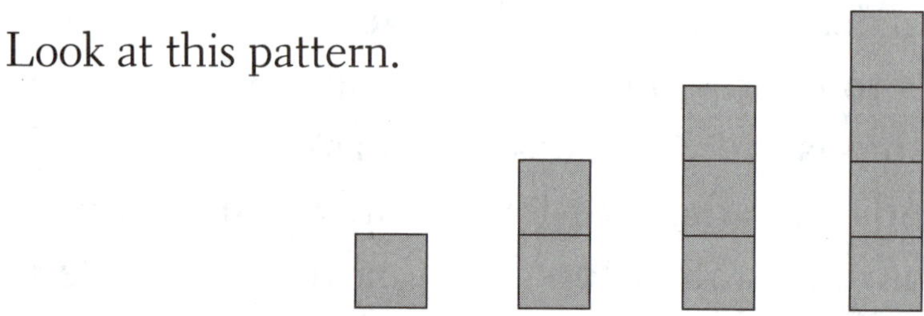

To learn how a pattern works, you can ask these questions:

A. **Does the pattern increase or decrease?**

The number of squares increases from 1 to 4.

B. **How much does the pattern increase or decrease each time?**

Each figure has 1 more square than the figure just before it.

C. **Describe the pattern. What is the rule?**

The rule is to add 1 square to the top of the last figure to get the next figure.

1. Look at this pattern of blocks.

Does the pattern increase or decrease? _______________________________

How much each time? _______________________________

Describe the pattern. What is the rule? _______________________________

____

Study each pattern. Then answer the questions.

2.  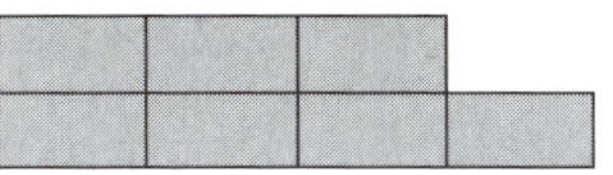 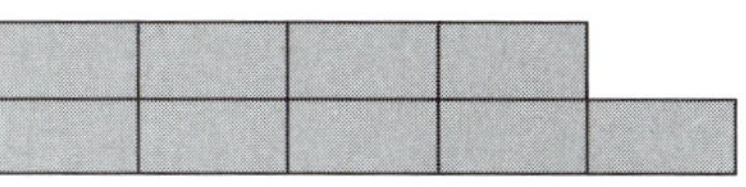

Does the pattern increase or decrease? _______________________

How much each time? _______________________

Describe the pattern. What is the rule? _______________________

____

3.

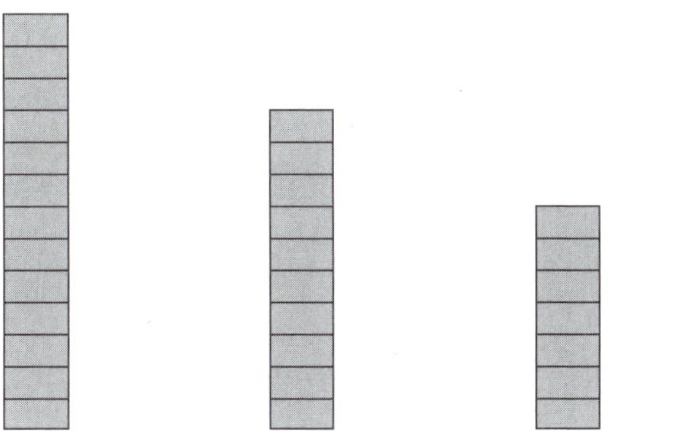

Does the pattern increase or decrease? _______________________

How much each time? _______________________

Describe the pattern. What is the rule? _______________________

____

4. Draw a pattern that increases. Write the rule.

It's a Fact!

The world's oldest jewelry may have been found in a cave in Africa. The 41 shells, which may have been used as beads, are more than 75,000 years old.

Rule: _______________________

Extending a Pattern

Some students in Mr. Arroyo's class decided to make
a necklace with seven strings of beads. The picture
below shows the first 4 strings. If they continue the
pattern, how many beads will be on strings 5, 6, and 7?

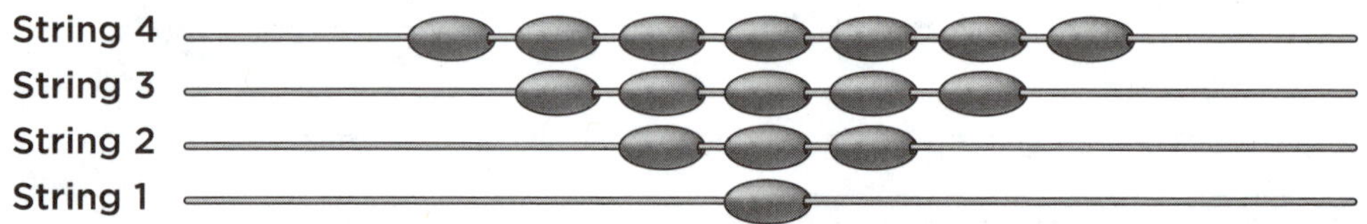

Follow the steps below to extend the pattern.

STEP 1 Study the pattern. Decide if it increases or decreases.

The number of beads increases from 1 to 7.

STEP 2 Tell by how much the pattern increases each time.

Each string has 2 more beads than the string just
before it.

STEP 3 Describe the pattern. Find the rule.

The rule is to add 2 beads to the last string
to get the next string of beads.

STEP 4 Use the pattern to solve the problem.

How many beads will be on string 5? ____________

How many beads will be on string 6? ____________

How many beads will be on string 7? ____________

Use the steps you learned to draw the next figure in each pattern.

1.

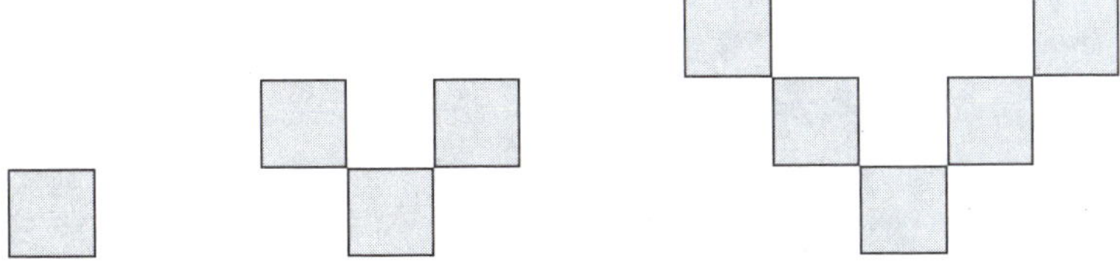

2.

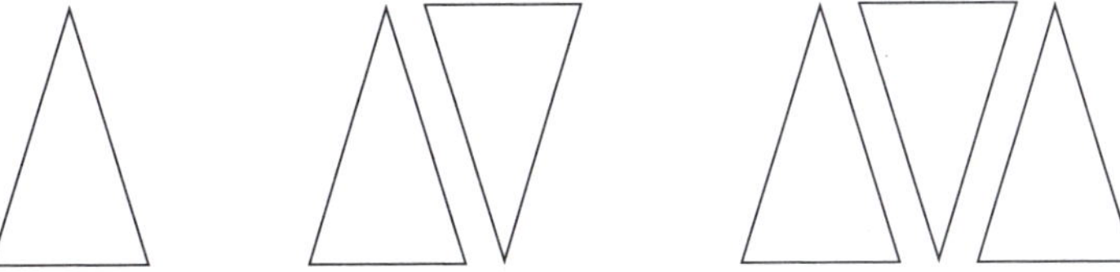

3.

On Your Own

Look at the first two patterns on this page. Draw the tenth figure in each pattern.

1. Look at the pattern below. Which shape comes next?

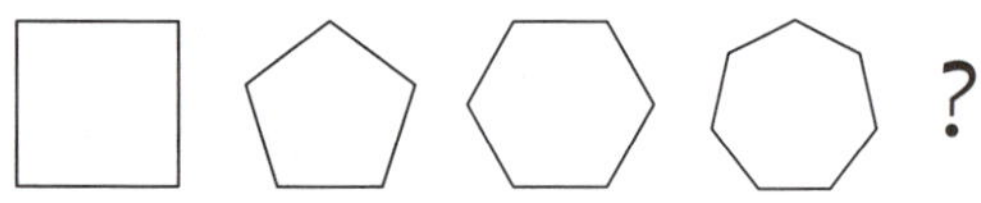

□△○□□□△○□□△ ?

Ⓐ rectangle

Ⓑ circle

Ⓒ square

Ⓓ triangle

2. Which shape comes next in this pattern?

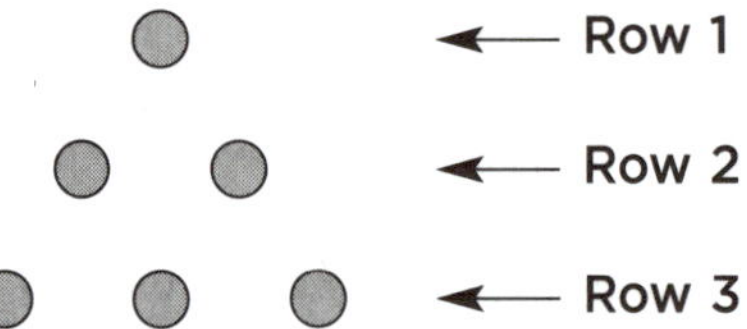

Ⓕ triangle   Ⓗ octagon

Ⓖ hexagon   Ⓙ decagon

3. Look at the pattern below. How many circles will be in the fifth row of the pattern?

● ← Row 1

● ● ← Row 2

● ● ● ← Row 3

Ⓐ 5

Ⓑ 6

Ⓒ 9

Ⓓ 15

4. What is the rule used for the pattern below?

```
                        X
                        X
                        X
                X       X
                X       X
                X       X
        X       X       X
        X       X       X
        X       X       X
X       X       X       X
```

Rule: _______________________________

____

5. Look at the pattern in Problem 4. How many X's will be in the seventh column of the pattern?

____

6. Think Back   Find the quotient.

$65 \div 2 =$ _______

Making Saving Fun

Press a frog's foot, watch its jaws open, and pop a coin in its mouth. That's what many children did in the late 1800s. While their parents put money in city banks, many children saved their coins inside iron roosters, cows, and pigs. Others owned whales, elephants, monkeys, and trick ponies. Companies made millions of these toy banks, making saving money fun.

Get Started

Tyrone earns money by doing chores for some of his neighbors. He saves money each week and puts it in his bank account. The chart below shows how much money he had in the bank at the end of each week.

Tyrone's Bank Account

Week 1	Week 2	Week 3	Week 4	Week 5
$4	$8	$12	$16	$20

How much did Tyrone save each week? $ ________________

Tyrone's sister Lisa does chores for neighbors too. The chart below shows how much money Lisa had in the bank at the end of each week.

Lisa's Bank Account

Week 1	Week 2	Week 3	Week 4	Week 5
$6	$12	$18	$24	$30

How much did Lisa save each week? $ ________________

Working with Functions

A **function** is a relationship between two sets of numbers.
If you know a number in one set, you can use a rule to
find the related number in the other set.

For example, suppose you want to trade nickels for pennies.
One nickel is worth 5 pennies, two nickels are worth 10
pennies, and so on. If you know how many nickels you have,
you can use a rule to find the number of pennies.

The table below shows this function.

To find the rule for this function,
look at each pair of numbers in
the table. What happens to the
number of nickels to make the
number of pennies?

Nickels	Pennies
1	5
2	10
3	15
4	20

If you know the number of nickels, you can multiply
by 5 to find the number of pennies. So the rule is $\times 5$.

Nickels	Rule	Pennies
1	$\times 5 =$	5
2	$\times 5 =$	10
3	$\times 5 =$	15
4	$\times 5 =$	20

1. How many pennies would you get for 5 nickels? _______________

2. How many pennies would you get for 10 nickels? _______________

Write the rule for each function shown below. The rule may use addition, subtraction, multiplication, or division.

3.

Pennies	Dimes
50	5
40	4
30	3
20	2

Rule: _______________

4.

Years	Months
1	12
2	24
3	36
4	48

Rule: _______________

5.

Yards	Feet
1	3
3	9
5	15
7	21

Rule: _______________

6.

Weeks	Days
1	7
2	14
3	21
4	28

Rule: _______________

7.

Pints	Quarts
6	3
8	4
10	5
12	6

Rule: _______________

8.

Michael's Age	Lily's Age
7	10
8	11
9	12
10	13

Rule: _______________

Solve a Problem

9. Maya saves $5 less each week than Sam. Complete the table to show how much Maya saves for each amount Sam saves.

Sam saves	Maya saves
$25	
$20	
$15	

Using Functions

Alissa opened a bank account so she could save money. She saved the same amount of money each week. The table below shows how much Alissa had in the bank at the end of 10 weeks. How much did Alissa have in the bank after 13 weeks?

Week	Dollars in the Bank
1	8
5	40
7	56
10	80
13	?

Follow the steps below to solve the problem.

STEP 1 Look for a rule that describes the relationship between the first pair of numbers.

$1 \times 8 = 8$ or $1 + 7 = 8$

STEP 2 Test the rule using the other pairs of numbers. The rule must work for all the other pairs.

$\times 8$	$+ 7$
$5 \times 8 = 40$	$5 + 7 = 40$
$7 \times 8 = 56$	
$10 \times 8 = 80$	

This equation is not true. So +7 cannot be the rule.

STEP 3 Use the rule to find the missing number.

$13 \times 8 =$ _____________

After 13 weeks, Alissa had $_____________.

Use the steps you learned to find the rule for the function shown in each table. Then complete the table.

1.

Decimeters	Centimeters
1	10
3	30
	50
7	
9	90

Rule: _______________

2.

Nickels	Quarters
30	6
25	
20	4
15	
10	2

Rule: _______________

3.

Sue's Age	Kim's Age
3	11
4	12
5	13
6	
7	

Rule: _______________

4.

Number of Hexagons	Number of Sides
2	12
4	24
	36
8	48
10	

Rule: _______________

On Your Own

Create your own function. Make a table to show your function. Then give your table to a friend and ask him or her to find the rule you used to make it.

16 Test Yourself

1. What is the rule for the function shown below?

Liam's Age	Tanesha's Age
10	5
11	6
12	7
13	8

Ⓐ ÷ 2 Ⓒ + 5

Ⓑ × 2 Ⓓ − 5

2. What is the rule for the function shown below?

Gallons	Quarts
2	8
4	16
6	24
8	32

Ⓕ + 6 Ⓗ × 4

Ⓖ − 6 Ⓙ ÷ 4

3. Which number is missing from the table?

Wheels	Tricycles
18	6
15	5
12	?
9	3

Ⓐ 2 Ⓒ 11

Ⓑ 4 Ⓓ 12

4. Complete the table below.

Feet	Inches
1	12
2	24
	36
4	

5. What is the rule for the function shown above?

Rule: _______________________

6. Think Back Find the product. Show your work.

$$\begin{array}{r} 25 \\ \times\ 9 \\ \hline \end{array}$$

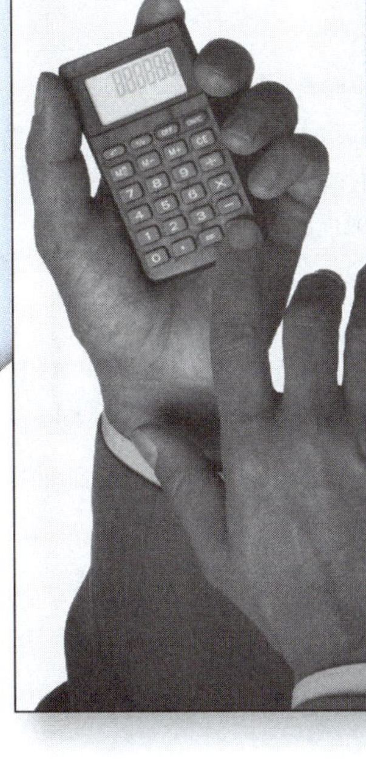

An Invention You Can Count On

Calculators are small tools that let you work with big numbers. Some are as thin as a credit card. But they haven't always been so small. An early calculator from 1909, for example, had a steel drum. It also had knobs, gears, and cranks. It was heavy and expensive. In today's dollars, this 23-pound calculator would cost $10,000.

Get Started

LaToya and Anna used a calculator to play a number game. LaToya added two one-digit numbers on the calculator and then showed the display to Anna. Then Anna named all the number sentences that LaToya might have used to get that sum.

In the space below each calculator, write four different number sentences LaToya might have used to get each sum shown in the calculator display.

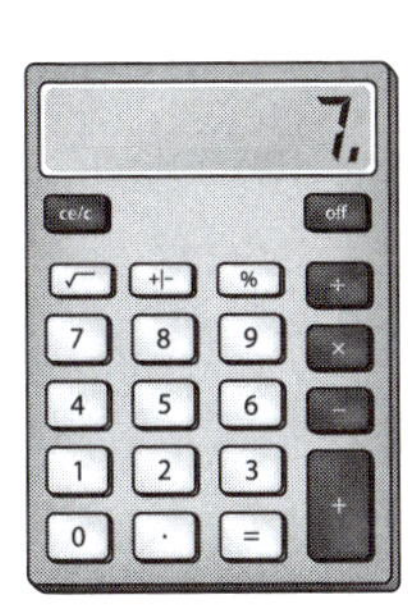

Working with Open Sentences

An **equation** is a number sentence that shows that two amounts are equal. An equation that has one of its numbers missing is sometimes called an **open sentence**:

$$5 + \square = 11$$

To find the missing number in an open sentence, you can use a fact triangle.

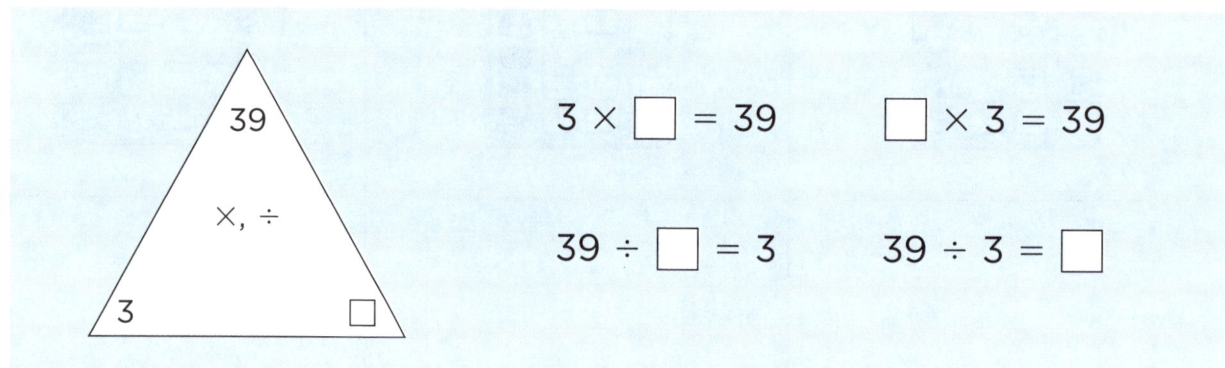

$$5 + \square = 11 \qquad \square + 5 = 11$$

$$11 - \square = 5 \qquad 11 - 5 = \square$$

The fact triangle above contains four open sentences. Look at the open sentences above. You can subtract to find the missing number: $11 - 5 = 6$.

You can also use a fact triangle to find the missing number in a multiplication or division equation.

$$3 \times \square = 39 \qquad \square \times 3 = 39$$

$$39 \div \square = 3 \qquad 39 \div 3 = \square$$

1. What is the missing number? _____________

Write the four open sentences for each fact triangle.
Then find the missing number.

2.

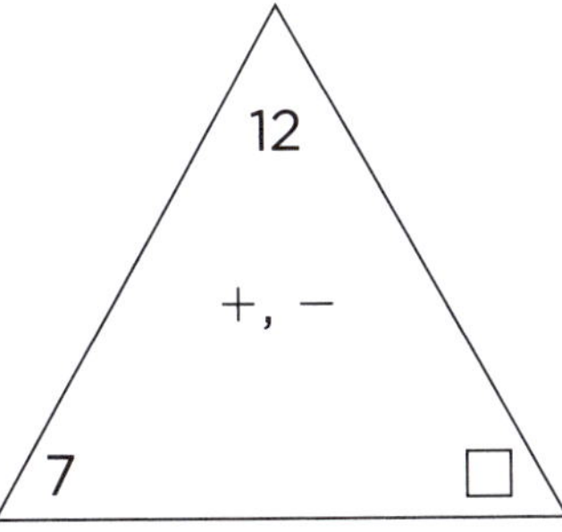

_______________________   _______________________

_______________________   _______________________

The missing number is _______.

3.

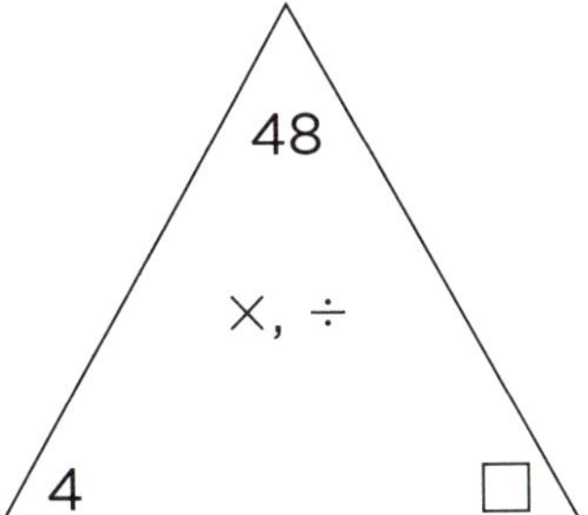

_______________________   _______________________

_______________________   _______________________

The missing number is _______.

Solve a Problem

4. Look at the fact triangle below.
Are the numbers related by addition
and subtraction or by multiplication
and division?

____

____

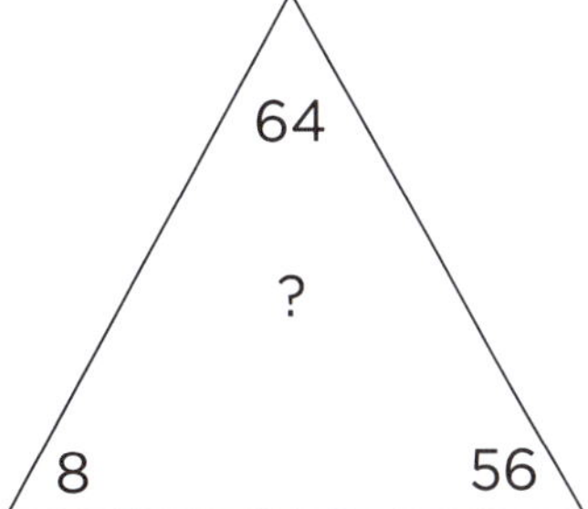

It's a Fact!

In the 1600s, a French mathematician named Blaise Pascal invented a mechanical calculating machine to help his father, a tax collector, add money amounts. It was called the Arithmetic Machine.

Solving Problems with Open Sentences

On Friday, Ms. Holt plans to review with her students how to use a calculator. She wants to have a calculator for each student in the class. Right now, Ms. Holt has 11 calculators. If there are 24 students in the class, how many more calculators does Ms. Holt need?

Follow the steps below to solve this problem.

STEP 1 Write the problem as an open sentence.

$$11 + \square = 24$$

number of calculators Ms. Holt has now

number of calculators needed

total number of students

STEP 2 Draw a fact triangle using the numbers from the open sentence. Write the four open sentences for the fact triangle.

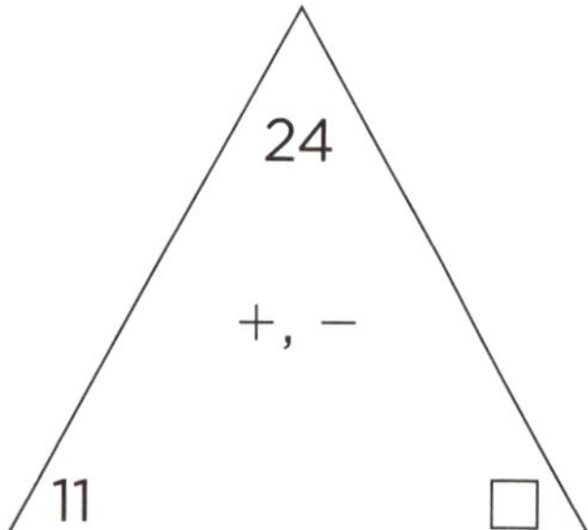

$11 + \square = 24$ _________

$24 - \square = 11$ _________

STEP 3 Use the open sentences to solve the problem.

Subtraction is the opposite of addition, so you can subtract to find the missing number.

$24 - 11 =$ _________

Follow the steps you learned to solve the problems below.

1. In Melissa's school, there are a total of 128 students in the fourth grade. If there are 67 girls in the fourth grade, how many boys are there?

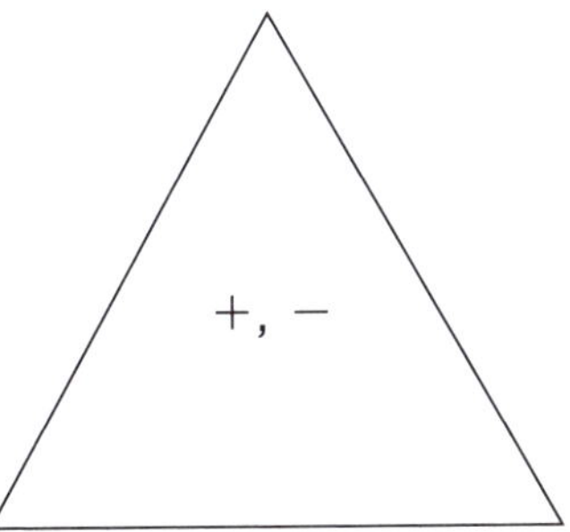

There are __________ boys in the fourth grade.

---

2. Brandon and Rosa are setting up chairs in the auditorium for a concert. The chairs need to be set up in 8 equal rows. If Brandon and Rosa use a total of 168 chairs, how many chairs will be in each row?

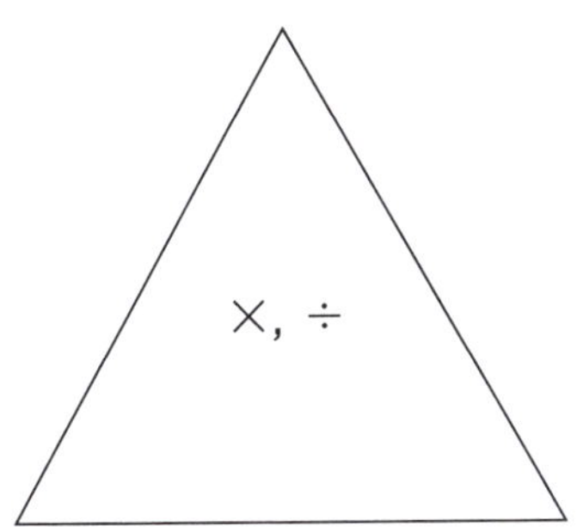

There will be __________ chairs in each row.

On Your Own

Find the missing numbers below. The same shape stands for the same number.

$$14 \times \square = \bigcirc$$
$$\bigcirc + 139 = 209$$

1. What number is missing from the fact triangle below?

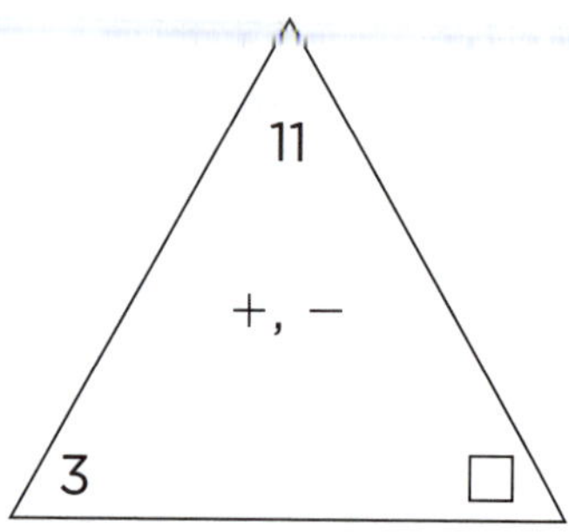

$$11$$
$$+, -$$
$$3 \qquad \square$$

Ⓐ 33   Ⓑ 14   Ⓒ 9   Ⓓ 8

2. What number goes in the box to make this equation true?

$$9 \times \square = 36$$

Ⓕ 3

Ⓖ 4

Ⓗ 9

Ⓙ 27

3. Solve the open sentence below.

$$15 - \square = 8$$

Ⓐ 23

Ⓑ 9

Ⓒ 8

Ⓓ 7

4. The sum of Kami's and Jamie's ages is 53. If Kami is 21, how old is Jamie?

____

5. What number goes in the box to make the equation true? Explain how you got your answer.

$$72 \div \square = 9$$

____

____

____

____

____

6. Think Back Divide. Show your work below.

$$95 \div 7 = \underline{\qquad}$$

It Doesn't Stick!

Imagine paying $82,500 for one penny! In 1996, someone did. The penny, stamped with the face of Abraham Lincoln, looks like a regular penny. However, all but 40 of the pennies made in 1943 were made of steel. A mistake at the mint, or coin factory, caused this penny to be made of copper. Use a magnet to check for a 1943 penny. If the penny doesn't stick, you may have a rare coin.

Get Started

Miguel and Shania collect pennies that are dated the year they were born. They compared their collections.

Miguel's Collection

Shania's Collection

- Miguel has ______ rows of pennies with ______ pennies in each row. Write a multiplication sentence to show how many pennies Miguel has. ______________________

- Shania has ______ rows of pennies with ______ pennies in each row. Write a multiplication sentence to show how many pennies Shania has. ______________________

Commutative Property

There are some things that are always true when adding or multiplying numbers. These are called **properties.**

The **Commutative Property of Addition** says that you can add numbers in any order, and the sum will always be the same.

Look at each sum below.

$$6 + 4 = 10 \qquad 4 + 6 = 10$$

The order of the addends does not change the sum.

The **Commutative Property of Multiplication** says that you can multiply numbers in any order, and the product will always be the same.

Look at each product below.

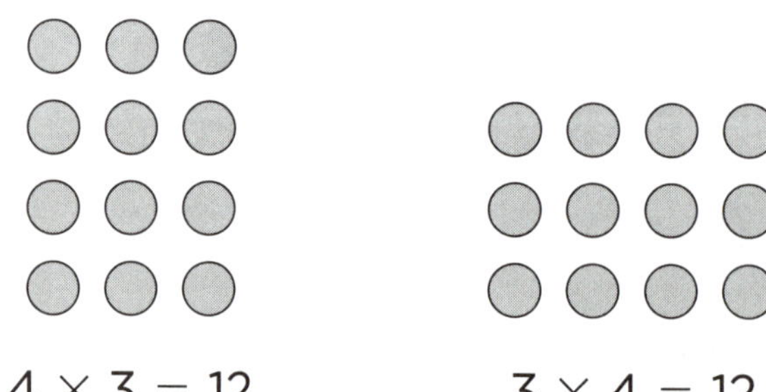

$$4 \times 3 = 12 \qquad 3 \times 4 = 12$$

The order of the factors does not change the product.

1. Find the missing factor: _____ $\times 17 = 17 \times 3$

2. Find the missing addend: $135 + 49 =$ _____ $+ 135$

Use the Commutative Property of Addition or Multiplication to rewrite the number sentences below.

3. $5 + 8 = 13$

_____ + _____ = _____

4. $8 \times 5 = 40$

_____ × _____ = _____

5. $19 + 5 = 24$

_____ + _____ = _____

6. $17 \times 9 = 153$

_____ × _____ = _____

Find each missing number below.

7. $48 \times 8 = $ _____ $\times 48$

8. _____ $+ 15 = 15 + 239$

9. $69 \times$ _____ $= 4 \times 69$

10. $45 + 15 = 15 +$ _____

Solve a Problem

11. Amanda says that since the commutative property works for multiplication, it also works for division, because division is the opposite of multiplication. Is she correct? Explain your answer.

____

____

____

____

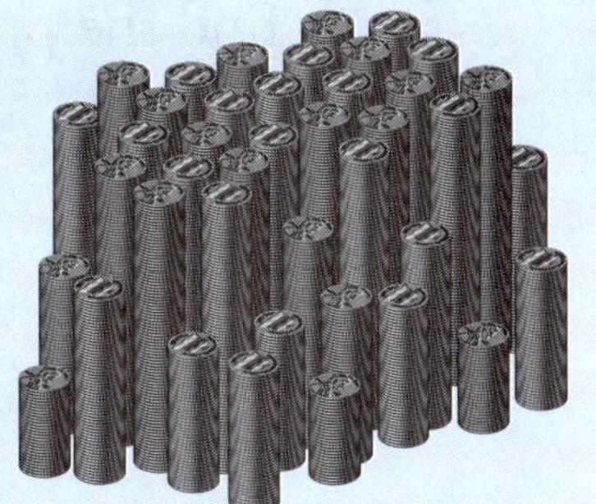

Associative Property

In addition or multiplication, numbers can be grouped
in any way and the answer will always be the same.
In mathematics, this is called the **associative property.**

Follow the steps below to see how the associative property
works for addition and multiplication.

STEP 1 Use parentheses to group the first two
numbers you will add or multiply. Do
what's inside parentheses first. Then
add or multiply from left to right.

$$(9 + 5) + 5 \qquad\qquad (3 \times 5) \times 2$$

$$14 + 5 = 19 \qquad\qquad 15 \times 2 = 30$$

STEP 2 Change the grouping by moving the
parentheses. Then add or multiply.

$$9 + (5 + 5) \qquad\qquad 3 \times (5 \times 2)$$

$$9 + 10 = \underline{\hspace{2em}} \qquad\qquad 3 \times 10 = \underline{\hspace{2em}}$$

Changing the grouping does not change the sum or the product.

Which grouping at the right makes it easier
to find the product? Explain your answer.

$$(3 \times 5) \times 2 = ?$$

$$3 \times (5 \times 2) = ?$$

____

____

____

Follow the steps you learned to use the Associative Property of Addition or Multiplication. The first one has been done for you.

1. $(7 + 8) + 3$

$\underline{\quad 15 + 3 = 18 \quad}$

$7 + (8 + 3)$

$\underline{\quad 7 + 11 = 18 \quad}$

2. $9 \times 4 \times 5$

$\underline{\qquad\qquad}$

$9 \times 4 \times 5$

$\underline{\qquad\qquad}$

3. $22 + 18 + 15$

$\underline{\qquad\qquad}$

$22 + 18 + 15$

$\underline{\qquad\qquad}$

4. $10 \times 3 \times 12$

$\underline{\qquad\qquad}$

$10 \times 3 \times 12$

$\underline{\qquad\qquad}$

Use the Associative Property of Addition or Multiplication to help you solve the problems below.

5. $3 \times 5 \times 2 = \underline{\qquad}$

6. $5 + 9 + 1 = \underline{\qquad}$

7. $6 + 4 + 13 = \underline{\qquad}$

8. $4 \times 5 \times 3 = \underline{\qquad}$

9. $7 \times 4 \times 25 = \underline{\qquad}$

On Your Own

Bill reached into his pocket and pulled out a quarter, a penny, two dimes, another quarter, a nickel, and then three more pennies.

Show how you can group and order Bill's coins to most easily find the total amount.

1. Which of the following shows an example of the Commutative Property of Multiplication?

Ⓐ $6 + 4 = 4 + 6$

Ⓑ $9 \times 3 = 3 \times 9$

Ⓒ $8 \times 0 = 0$

Ⓓ $8 \times 1 = 8$

2. Which of the following shows an example of the Associative Property?

Ⓕ $16 + 0 = 16$

Ⓖ $2 \times 10 = 10 \times 2$

Ⓗ $5 \times (3 + 2) = (5 \times 3) + (5 \times 2)$

Ⓙ $(4 + 7) + 3 = 4 + (7 + 3)$

3. Which number makes this equation true?

$$5 \times (4 \times 3) = (5 \times \boxed{}) \times 3$$

Ⓐ 3

Ⓑ 4

Ⓒ 15

Ⓓ 20

4. Name the property used for each number sentence below.

$$8 + 5 = 5 + 8$$

Property: _______________

$$(3 \times 8) \times 10 = 3 \times (8 \times 10)$$

Property: _______________

5. Circle the operation(s) below where the Commutative Property will **<u>not</u>** work? Explain your thinking.

Addition
Subtraction
Multiplication
Division

____

____

____

____

6. Think Back Compare the fractions below using >, <, or =.

$$\frac{3}{9} \bigcirc \frac{2}{3}$$

Our Nation's Library

The Library of Congress in Washington, D.C., is the largest library in the world. It has more than 29 million books and other printed materials. It also has almost 5 million maps, 12 million photographs, and 3 million recordings. The library's materials are stored on almost 530 miles of bookshelves. That's longer than the Potomac River, which flows through the nation's capital not far from the library.

Get Started

Students in Mr. Martin's class are bringing books that they have read to add to their school library. The table on the right shows how many books some students are bringing in.

Name	Number of Books
Sue	8
Juanita	4
Carlos	10
Tom	6

- Sue and Juanita will carry their books to the library. How many books will each girl carry if they each carry the same amount?

__________ books each

Explain your answer on the lines below.

____

____

- Carlos and Tom will also carry their books to the library. How many books will each boy carry if they each carry the same amount?

__________ books each

Explain your answer on the lines below.

____

____

Working with the Mean

On page 113, you made equal groups to answer
the questions. You can use a picture to help you
make equal groups.

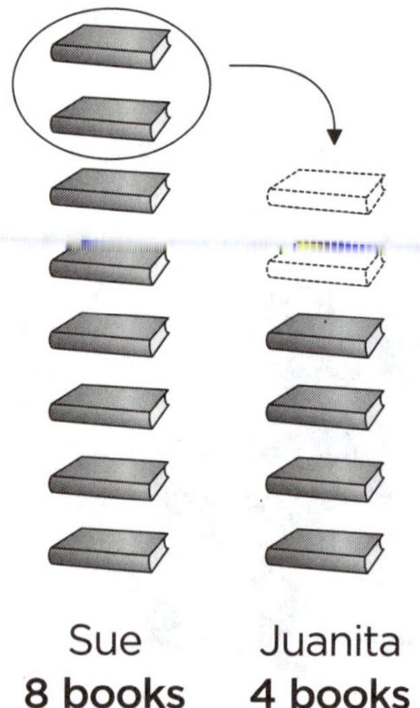

The picture on the right shows the number
of books Sue and Juanita will carry to
the library.

If Sue gives 2 books to Juanita, they
will each carry 6 books to the library.

When you even out the numbers in a group of data so that
all the numbers are the same, you are finding the **mean,**
or average.

So the mean of 8 and 4 is 6. You can say that Sue and Juanita
will carry an average of 6 books each.

This picture shows the number of books
Carlos and Tom will carry to the library.

If Carlos gives Tom 2 books, they will
each carry 8 books to the library.

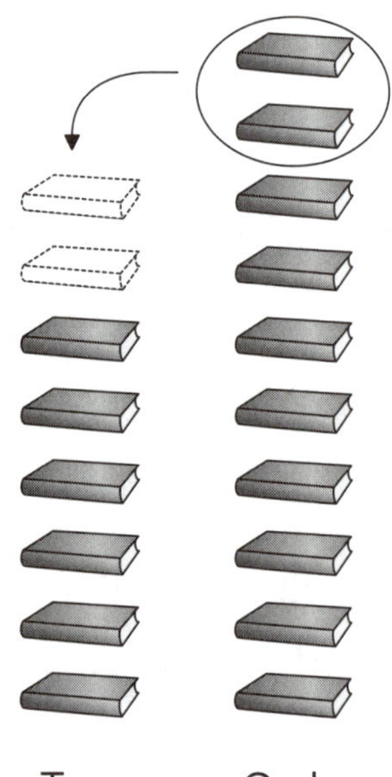

1. So the mean of 6 and 10 is __________ .

2. Carlos and Tom will carry an average of

   __________ books each.

Use the pictures to find the mean, or average.

3. The picture shows the number of quarters Abigail and Sandy collected.

Abigail
7 quarters

Sandy
3 quarters

The mean of 7 and 3 is __________.

Abigail and Sandy collected an average of __________ quarters each.

4. The picture shows the number of books Robert and Melinda read over the summer.

Robert
12 books

Melinda
6 books

The mean of 6 and 12 is __________.

Robert and Melinda read an average of __________ books each.

Solve a Problem

5. Suppose Jay read 3 books over the summer. If you added this data to the data in Problem 4, what would the new mean be?

Mean: __________

It's a Fact!
The Library of Congress has the largest collection of comic books in the world.

Find the Mean

Another way to find the mean, or average, is to use addition and division.

The table at the right shows the data from page 113.

Suppose you want to find the average number of books each student will carry to the library. You can follow the steps below.

Name	Number of Books
Sue	8
Juanita	4
Carlos	10
Tom	6

STEP 1 Find the sum of the numbers in the data.

$8 + 4 + 10 + 6 =$ _________

STEP 2 Count the number of addends.

$8 + 4 + 10 + 6$
  ↑   ↑   ↑   ↑
  1   2   3   4

Remember
Addends are the numbers you add in an addition problem.

There are 4 addends.

STEP 3 Divide the sum in Step 1 by the number of addends you counted in Step 2. The answer is the mean.

$28 \div 4 =$ _________
mean

The four students will carry an average of _________ books each.

Use the steps you learned to find the mean of the data in each group. Show your work.

1.

Name	Books Read
Corinne	35
Michael	40
Ryan	24

The mean is __________.

2.

Name	Age
Jen	13
Sarah	13
Rosalva	16
Vicki	14

The mean is __________.

3.

Day	Temperature
Monday	25°C
Tuesday	33°C
Wednesday	38°C

The mean is __________ °C.

On Your Own

Use addition and division to solve the problems on page 115. Explain how this method of finding the mean works.

1. Carlos got an 88 on his first math quiz and a 92 on his second math quiz. What was his average score?

Ⓐ 92

Ⓑ 90

Ⓒ 88

Ⓓ 86

2. Helen bowled three games. She scored 135, 60, and 105. What was Helen's average bowling score?

Ⓕ 60

Ⓖ 96

Ⓗ 100

Ⓙ 105

3. Four students read an average of 25 books each. How many books did they read in all?

Ⓐ 21

Ⓑ 29

Ⓒ 50

Ⓓ 100

4. The table below shows the heights of six fourth-grade students. What is the mean of the data?

Name	Height
Pat	52 in.
Kay	47 in.
Joan	50 in.
Lucy	52 in.
Whitney	51 in.
Dionne	48 in.

Mean: ________ in.

5. Explain the steps you used to find your answer to Question 4.

____

____

____

____

6. Think Back Find the quotient. Show your work.

$$37 \div 3 = \underline{\qquad}$$

Weather

How often do you watch the clouds in the sky? Have you ever seen a scud? What about a mare's tail or an anvil? Did you know that these are kinds of clouds? Did you also know that people who study the weather use clouds to help them predict the weather? The next time you're outside, look up. If you see fractus clouds, it may be time to open your umbrella.

Get Started

Ling's class studied temperature changes in their state for one year. They recorded the average monthly temperature of every other month from January to November. The table below shows the data they collected. Use the table to answer the questions below.

Average Monthly Temperatures

Month	Average Temperature
January	30°F
March	35°F
May	60°F
July	70°F
September	65°F
November	50°F

- Which month had the lowest average temperature? _______________

- How much did the average temperature decrease from September to November? _______________ °F

- How much did the average temperature increase from January to May? _______________ °F

Working with Line Graphs

You can use a line graph when you want to show how something changes over time. A line graph uses line segments and points to show data.

The line graph below shows the data from page 119.

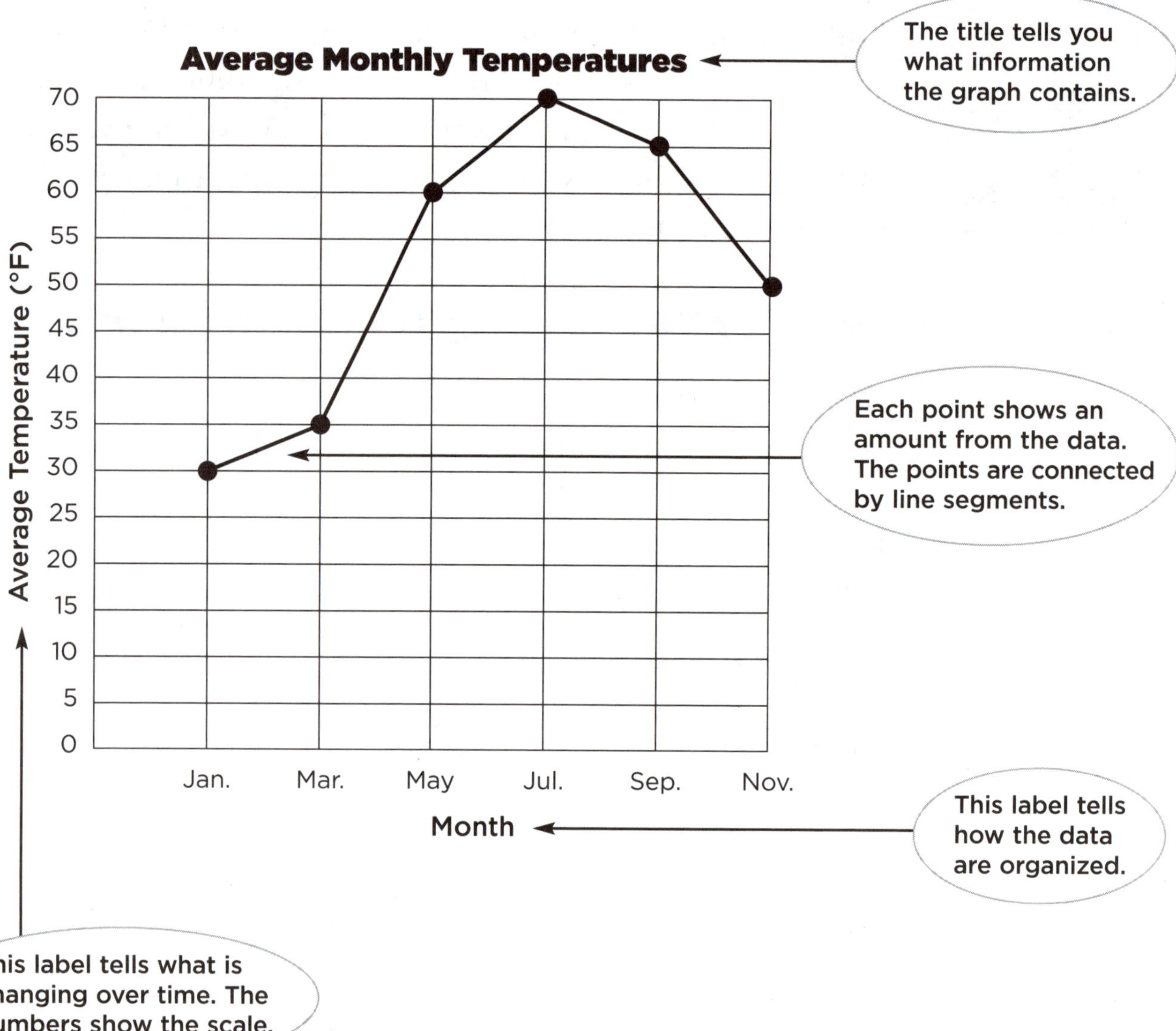

1. In which two-month period was there the greatest increase in average temperature? Explain how you can tell by looking at the graph.

____

____

Jeremy wanted to find out when the traffic was the busiest in front of his house. He did this by counting the number of cars that were stopped at the traffic light on the corner. He counted cars once each hour from 8 A.M. to 3 P.M. The line graph below shows the data Jeremy collected. Use the line graph for Questions 2–4.

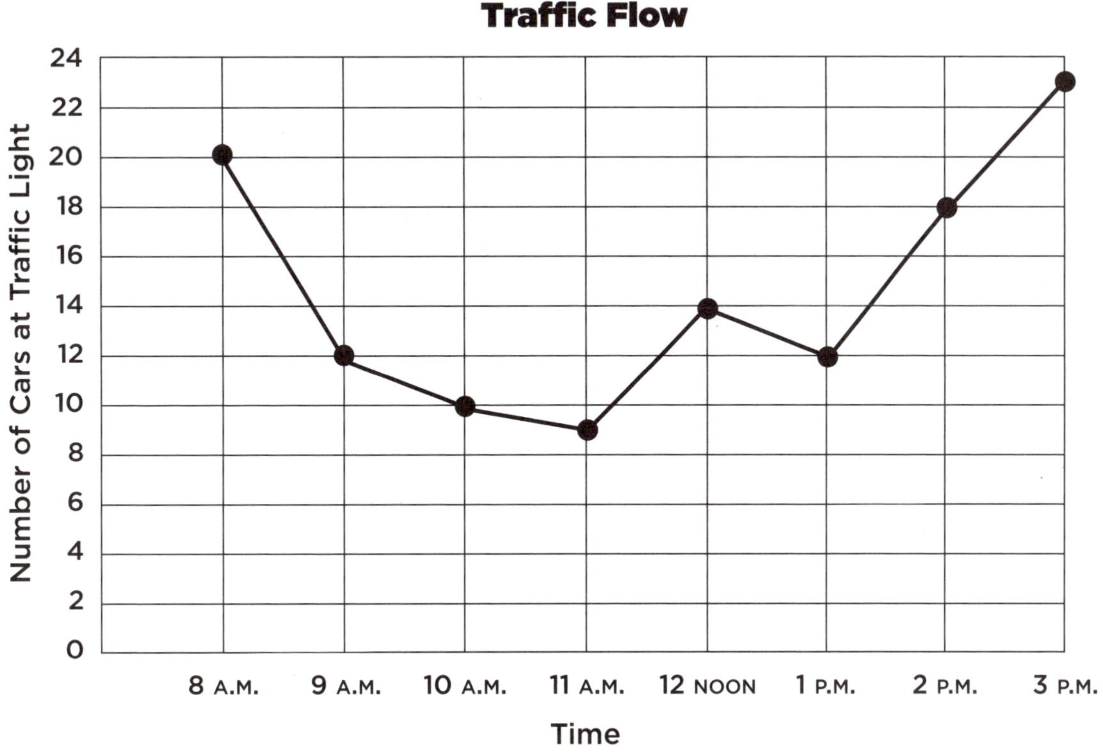

2. At what time was the traffic the busiest? _________________

3. At what time was the traffic the least busy? _________________

4. Why do you think there was a large decrease in traffic between 8 A.M. and 9 A.M.?

____

____

Constructing a Line Graph

Caytie recorded these daily high temperatures in her city for one week in March: 40°F, 31°F, 30°F, 37°F, 45°F, 55°F, and 40°F. Now she wants to make a line graph to show her data.

Use the blank graph on the next page and follow the steps below to help Caytie make a line graph.

STEP 1 Organize the data in a table.

Complete this table using the data above.

Day	High Temperature (°F)
Day 1	
Day 2	
Day 3	
Day 4	
Day 5	
Day 6	
Day 7	

STEP 2 Choose a scale for the graph.

If possible, use a scale that includes most of the numbers in the data.

> **Remember**
> The scale always starts at 0.

STEP 3 Write the labels and give the graph a title.

STEP 4 Draw points on the graph to show the data.

STEP 5 Connect the points with straight line segments.

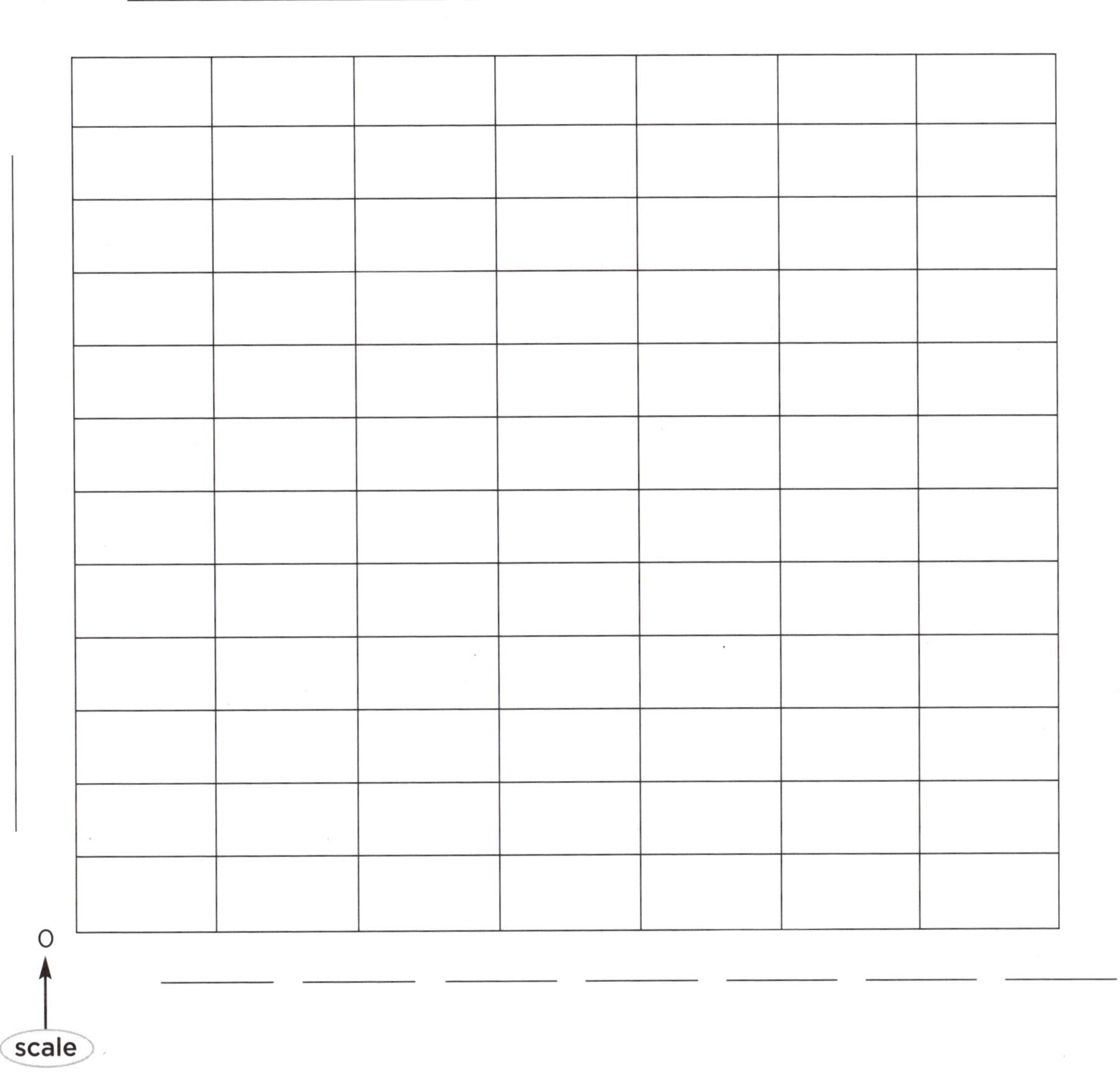

1. In the line graph above, what is changing over time?

____

____

On Your Own

Ask a partner how he or she chose the scale for the line graph above.

Explain how you chose the names for the labels.

20 Test Yourself

Use the line graph below to answer Questions 1–5.

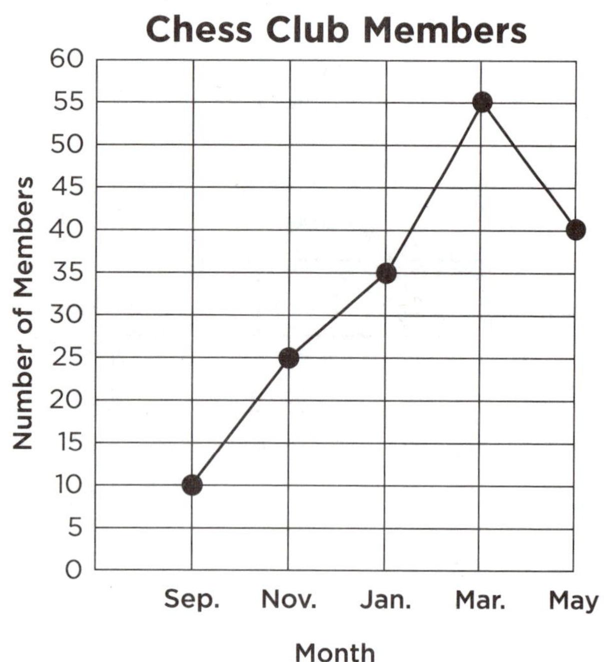

1. How many members were in the chess club in May?

Ⓐ 25   Ⓒ 40

Ⓑ 35   Ⓓ 55

2. Between which two months did the number of chess club members decrease?

Ⓕ September and November

Ⓖ March and May

Ⓗ November and January

Ⓙ January and March

3. Between which two months did the number of members increase the most?

Ⓐ September and November

Ⓑ November and January

Ⓒ January and March

Ⓓ March and May

4. How many more members were in the chess club in May than in September?

____

5. What conclusions can you make from the line graph?

____

____

____

____

6. Think Back Draw the next figure in the pattern below.

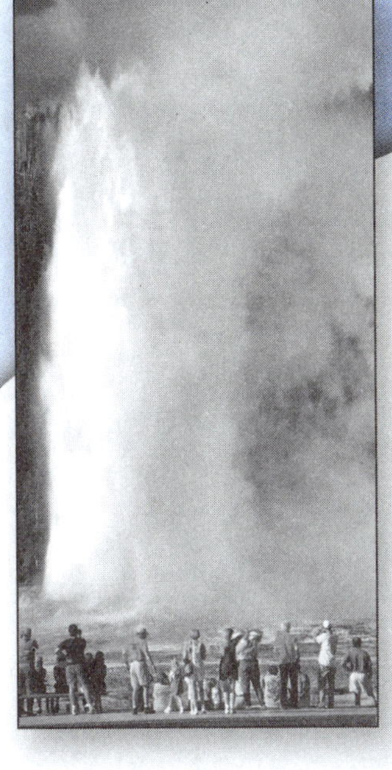

Old Faithful

Heat from the earth warms water trapped beneath the surface. The water boils and gases rise. Suddenly, as many as 8,400 gallons of boiling water explode from the ground. For minutes, the steaming fountain bubbles upward as high as 184 feet. Then it stops, settling quietly until its next performance about 91 minutes later. This predictable geyser is called Old Faithful, and it attracts almost 3 million visitors to Yellowstone National Park each year.

Get Started

Talia's family was planning a vacation to one of three national parks—Yellowstone, Yosemite, or Zion. Each family member voted for one park. They wrote their choices on separate pieces of paper and put them in a bag. Here are their votes.

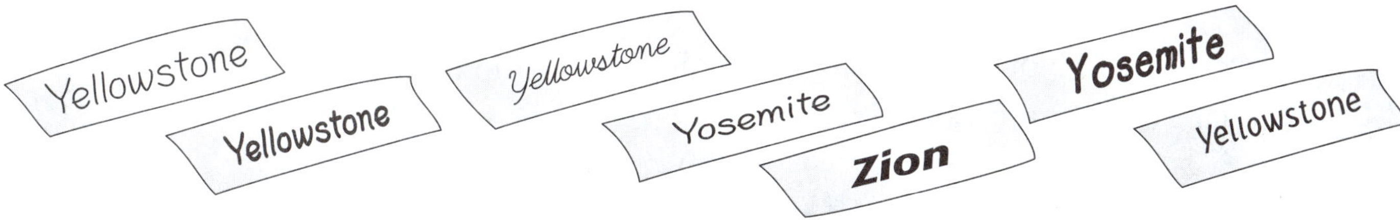

The family chose Talia to choose the park. She closed her eyes and took one piece of paper out of the bag.

- Which national park is Talia most likely to choose? _______________

Explain your answer.

____

____

____

Probability

Probability is the chance, or likelihood, that an **event** will happen. You can use words such as *likely, unlikely, certain,* and *impossible* to describe the probability of an event.

Look at the spinner below.

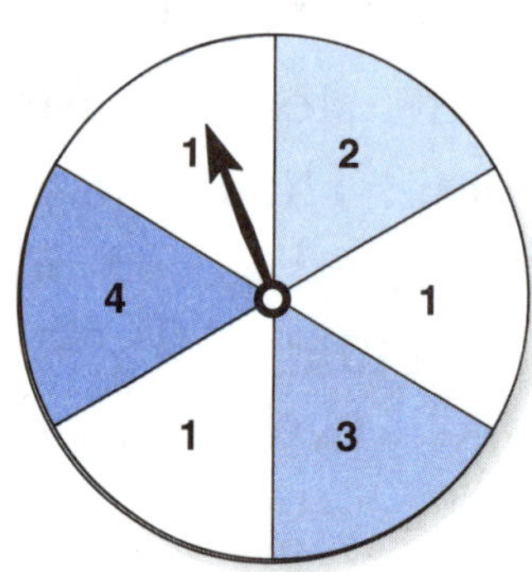

The spinner has six equal sections. Imagine spinning the spinner one time. There are six possible outcomes.

What is the chance that the pointer will stop on 4? We use the term favorable outcome to describe the outcome we are looking for. So in this example, landing on 4 is a favorable outcome.

There is a 1 in 6 probability that the pointer will stop on 4. You can write this probability as a fraction.

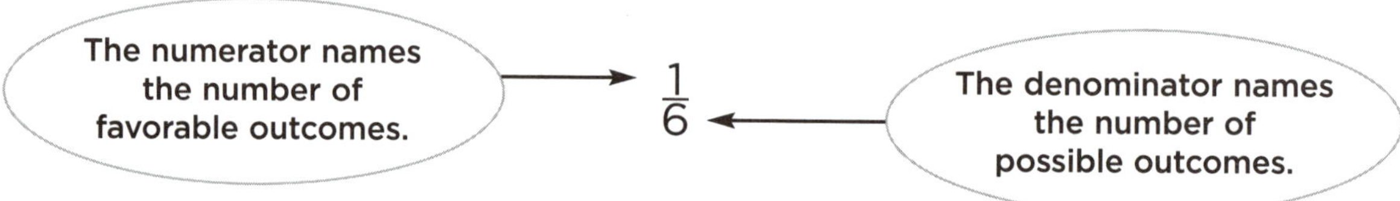

1. What is the probability that the pointer will land on 3?

Write your answer as a fraction. _________

2. What is the probability that the pointer will land on 1?

Write your answer as a fraction. _________

Imagine spinning the spinner to answer each question below.

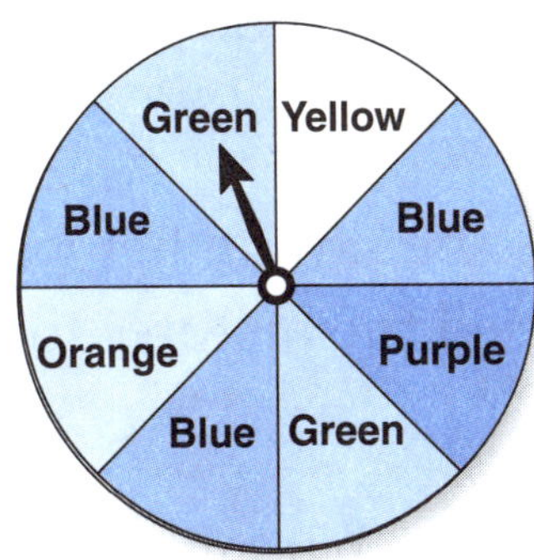

3. What is the probability of landing on Orange?
Write your answer as a fraction: _______

Explain your answer. _______________________

____

4. What is the probability of landing on Blue?
Write your answer as a fraction: _______

Explain your answer. _______________________

____

5. What is the probability of landing on Green?
Write your answer as a fraction: _______

Explain your answer. _______________________

____

6. Look at the spinner above. Use words to describe
the probability of landing on Red.

____

____

It's a Fact!
Two thirds of all
of the world's
geysers are
in Yellowstone
National Park.

Predicting and Testing Outcomes

You can use probability to predict an outcome of an experiment. You can also test your prediction by doing the experiment.

Talia drew pictures of the animals she saw on her visit to Yellowstone National Park. She used the pictures to make a set of playing cards.

Suppose you put the playing cards in a hat and pick one without looking. What is the probability of picking each animal? Follow the steps below to make predictions.

Coyote	Elk	Chipmunk	Chipmunk	Lynx	Lynx	Lynx	Bald Eagle	Bald Eagle	Bald Eagle

STEP 1   Count the number of possible outcomes. You will pick one card without looking. There are 10 cards, so there are 10 possible outcomes.

STEP 2   Count the number of favorable outcomes for each animal. For example, there is one Coyote card, so there is one favorable outcome for Coyote.

Coyote: _____1_____   Elk: _________

Chipmunk: _________   Lynx: _________

Bald Eagle: _________

STEP 3   Use fractions to describe the probability of each outcome.

Coyote: $\frac{1}{10}$ Elk: $\frac{1}{10}$ Chipmunk: $\frac{2}{10}$
Lynx: $\frac{3}{10}$ Bald eagle: $\frac{3}{10}$

Do an experiment to test your prediction. Shuffle the cards your teacher gives you and place them facedown on a table. Without looking, draw one card, record the outcome in the tally table below, and return the card to the deck. Once again, shuffle the cards and place them facedown. Then draw another card and repeat the same steps until you have drawn a total of ten times.

Animal Name on Card	Tally
Bald Eagle	
Chipmunk	
Coyote	
Elk	
Lynx	

1. Write fractions to explain the outcomes of your experiment.

   Bald eagle: _________   Elk: _________

   Chipmunk: _________   Lynx: _________

   Coyote: _________

2. Compare your predictions to the results of your experiment. Were they the same or different? Explain your answer.

   ____

   ____

3. What do you think would happen if you drew ten more times? Explain your thinking.

   ____

   ____

On Your Own

Find a spinner from a game you have at home or in the classroom. Predict the probability of each outcome. Then do an experiment to test your predictions.

 Test Yourself

Use this spinner to answer
Questions 1–4.

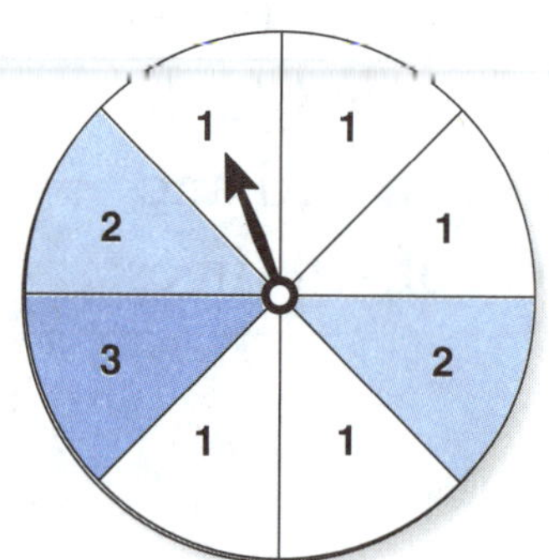

1. What is the probability, expressed
   as a fraction, that the pointer on
   the spinner will stop on 1?

   Ⓐ $\frac{1}{8}$ Ⓒ $\frac{5}{8}$

   Ⓑ $\frac{2}{8}$ Ⓓ $\frac{8}{8}$, or 1

2. What is the probability, expressed
   as a fraction, that the pointer on
   the spinner will stop on 3?

   Ⓕ $\frac{1}{8}$ Ⓗ $\frac{3}{8}$

   Ⓖ $\frac{2}{8}$ Ⓙ $\frac{4}{8}$

3. What word describes the
   probability of landing on 1 with
   one spin?

   Ⓐ Likely

   Ⓑ Unlikely

   Ⓒ Certain

   Ⓓ Impossible

4. What word describes the
   probability of landing on 1, 2, or 3
   with one spin?

   Ⓕ Certain

   Ⓖ Likely

   Ⓗ Unlikely

   Ⓙ Impossible

5. Imagine that the number
   cards below are placed facedown.
   If you draw one card without
   looking, which card are you
   most likely to choose?

6. **Think Back** Write a fraction
   below that is equal to 1 whole.

   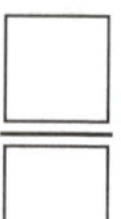

Clowning Around

Paint your face and put on a squeaky nose. Wear a polka-dotted suit, a matching hat, and a wig. Step into floppy shoes, and you're ready to learn—to be a clown. At clown schools, people learn to juggle, walk on stilts, and ride unicycles. They tumble like acrobats. They twist balloons into animals and do tricks. They walk on tightropes and zoom out of cannons. Clowning around is hard work.

Get Started

Corky the Clown is going to entertain at a party. One way Corky makes the children laugh is by wearing 3 hats at the same time. He has a red hat (R), a green hat (G), and a yellow hat (Y).

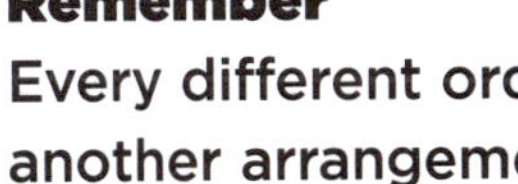

- How many different ways can Corky place these hats on his head? One way is shown above. Show the other ways Corky can wear these hats.

> **Remember**
> Every different order is another arrangement.

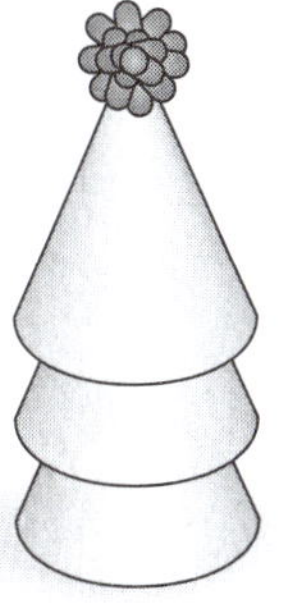
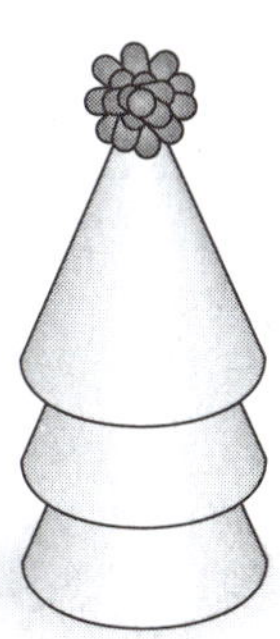
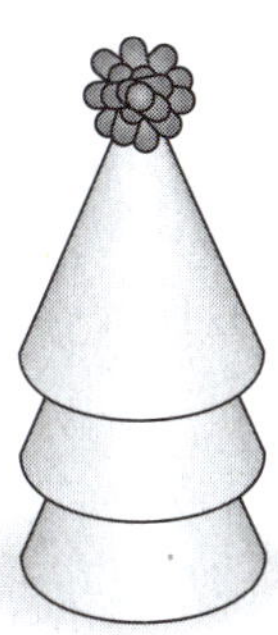
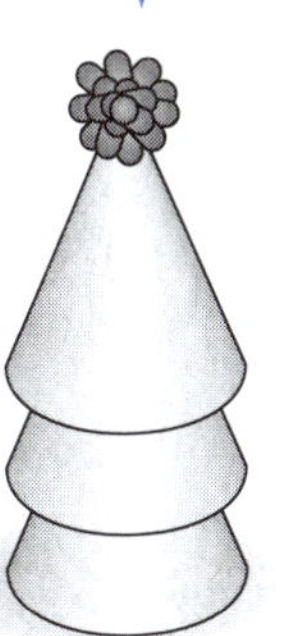
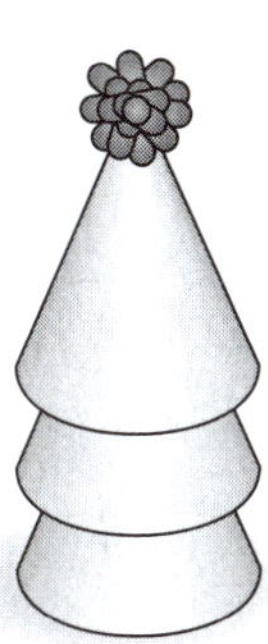

Working with Combinations

You saw that there are 6 different ways to arrange Corky's three hats. When you make arrangements, each group is different because no group has the exact same order as any other group.

When you make **combinations**, each group is different because no group has the exact same <u>items</u> as any other group.

The picture at the right shows the shoes that Corky can wear to the circus. He can pick two shoes to wear.

Complete the list below to show the combinations Corky has to choose from. Remember, no group can have the exact same items as any other group.

Flowers, Dots	Flowers, _________	Flowers, Stripes
~~Dots, Flowers~~	Dots, _________	Dots, _________
		Stars, _________

1. How many combinations are there in all? _________

2. Why is the pair Stars, Flowers not listed above?

____

____

Use what you have learned to solve each problem below.

3. Lou, Jose, Aisha, and Frances are helping to set up for their school carnival. Two of the students are needed to set up the Ring Toss booth. List the different combinations of two students that are possible.

__________ , __________   __________ , __________

__________ , __________   __________ , __________

__________ , __________   __________ , __________

4. The menu at the right shows the different flavors of frozen yogurt you can buy at the carnival. Suppose you want two scoops. List the different combinations of two scoops that are possible.

__________ , __________   __________ , __________

__________ , __________   __________ , __________

__________ , __________   __________ , __________

__________ , __________   __________ , __________

__________ , __________

Solve a Problem

5. Look back at page 131. Suppose Corky wears two hats at the same time. How many combinations of two hats are possible?

____

It's a Fact!

Canoes, lollipop toes, saddles, checkers, Santa boots, sneakers, tramps, and jesters are different kinds of clown shoes.

Combinations and Tree Diagrams

Maribel the Clown has two shirts—red and yellow—and
two pairs of pants—blue and green. How many different
shirt-and-pants combinations can she make?

One way to solve this problem is to use a tree diagram.
Follow the steps below.

STEP 1 List the names of the shirts.

<u>Shirts</u>

Red

Yellow

STEP 2 Use a tree diagram to match each
shirt with the two choices of pants.

<u>Shirts</u>   <u>Pants</u>

Red < Blue / Green

Yellow < Blue / Green

STEP 3 List all the possible combinations.

<u>Shirts</u>   <u>Pants</u>

Red < Blue → Red shirt, _______ pants
Green → Red shirt, _______ pants

Yellow < Blue → _______ shirt, _______ pants
Green → _______ shirt, _______ pants

Follow the steps you learned to solve the problem below.

1. Suppose Maribel the Clown has another shirt to choose from. This shirt is purple. Now how many different shirt-and-pants combinations can Maribel make? Complete the tree diagram below.

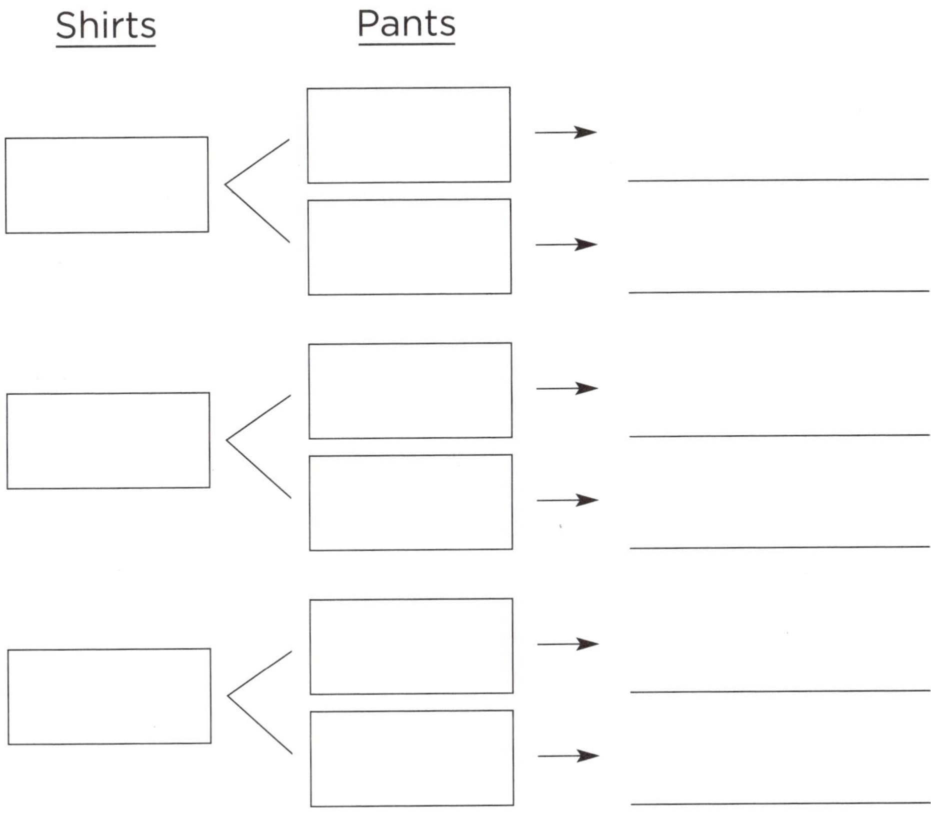

2. Jayse says that you can find the number of shirt-and-pants combinations above by multiplying the number of shirt choices by the number of pants choices. Is Jayse right? Explain.

____

____

____

____

On Your Own

Make tree diagrams for Problems 3 and 4 on page 133.

Compare your tree diagrams to the lists you made.

1. Hector has 4 different shirts and 2 different pairs of pants. How many different outfits can he make?

  Ⓐ 2   Ⓑ 6   Ⓒ 8   Ⓓ 12

2. How many different combinations does this tree diagram show?

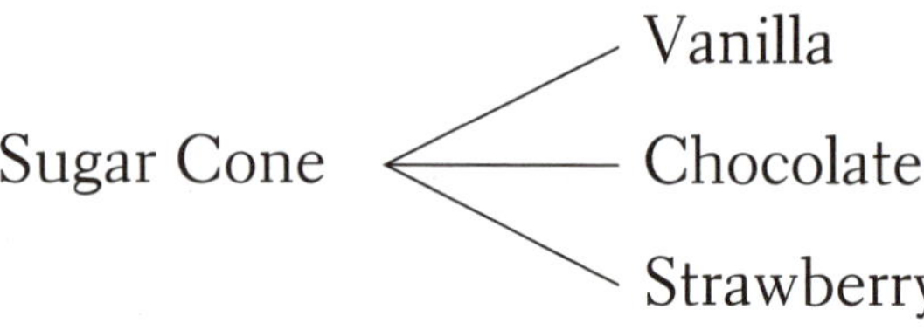

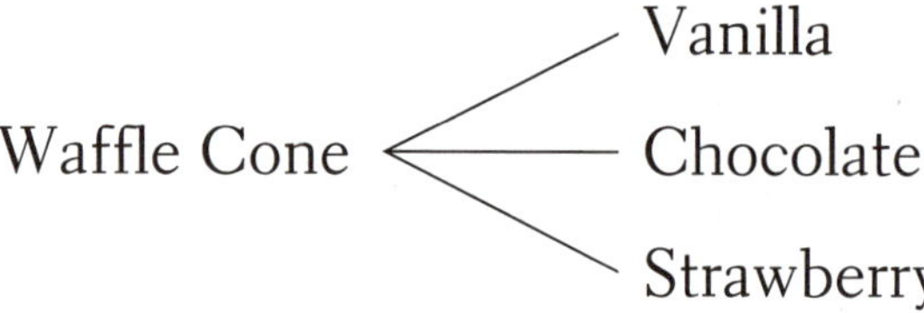

  Ⓕ 2   Ⓖ 6   Ⓗ 8   Ⓙ 9

3. How many different combinations of two letters can be made from the word below?

FOURTH

  Ⓐ 30

  Ⓑ 15

  Ⓒ 6

  Ⓓ 5

4. Use the menu below. Draw a tree diagram to show the possible sandwich-and-drink combinations.

5. What is the total number of sandwich-and-drink combinations?

____

6. Think Back Multiply.

$$3 \times 24 = \underline{\hspace{2cm}}$$

Glossary

acute angle the rays of an acute angle form an opening that is narrower than that of a right angle

angle a geometric figure formed by two rays that share the same endpoint

area the number of square units that cover a figure without overlapping. The area of a rectangle can be found by multiplying its length and width

arrangement a special way of ordering a collection of things

associative property when adding or multiplying, numbers can be grouped in any way and the sum or product will always be the same

clockwise in the direction that the hands on a clock move

combination a grouping of different items. When you make combinations, each group is different because no group has the exact same items as any other group

Commutative Property of Addition numbers can be added in any order, and the sum will always be the same

Commutative Property of Multiplication numbers can be multiplied in any order, and the product will always be the same

counterclockwise in the opposite direction that the hands on a clock move

decimal a number with a decimal point

decimal point a dot (.) used to separate the ones place and the tenths place in a decimal

denominator the bottom number in a fraction

endpoint a point at each end of a line segment or at the beginning of a ray

equation a number sentence that shows that two amounts are equal

equivalent fractions fractions that name the same part of a whole. For example, $\frac{1}{2}$ and $\frac{2}{4}$ are equivalent fractions

favorable outcome an outcome you are looking for in a probability experiment

function a relationship between two sets of numbers. If you know a number in one set, you can use a rule to find the related number in the other set

hundredths the hundredths place is to the right of the tenths place in a decimal

improper fraction a fraction with a numerator greater than or equal to its denominator

intersecting lines two or more lines that cross

line a straight path that goes on forever in two directions

line segment a straight path connecting two points

mean a number that describes a group of data. To find the mean, add all the numbers in the data and divide by the number of addends

mixed number a number with a whole number part and a fractional part

numerator the top number in a fraction

obtuse angle the rays of an obtuse angle form an opening that is wider than that of a right angle

open sentence an equation with one of its numbers missing

outcome something that could happen in a probability experiment

parallel lines lines that are always the same distance apart and never intersect

parallelogram a quadrilateral whose opposite sides are parallel

partial products the results when multiplying the ones, then the tens, and so on, in a multiplication problem

perimeter the distance around the outside of a figure

perpendicular lines intersecting lines that form square corners, or right angles

point in geometry, a point is an exact location. You can show a point by drawing a dot

point of intersection the point at which two or more lines cross

polygon a closed figure whose sides are all line segments

probability the chance, or likelihood, that an event will happen properties things that are always true about numbers

quadrilateral a polygon with four sides and four angles

ray a geometric figure with one endpoint that goes on forever in one direction

rectangle a parallelogram with 4 right angles

reflection flipping a figure over a line

remainder the amount left over after dividing

rhombus a parallelogram whose sides are all equal in length

right angle an angle whose rays form a square corner

rotation turning a figure around a point

square a polygon with opposite sides that are parallel, 4 right angles, and 4 sides that are all the same length. A square is a parallelogram, a rectangle, and a rhombus

tenths the tenths place is to the right of the decimal point in a decimal

trapezoid a quadrilateral with exactly one pair of parallel sides

vertex in an angle, the vertex is the endpoint shared by the two rays

yard a measurement equal to 3 feet, or 36 inches

Number Line

A number line can help you add and subtract. To add
2 and 4, put your finger on the 2. Then move it to the right 4
spaces. Your finger should end up on 6, so you know that 2
+ 4 = 6.

To subtract 3 from 5, put your finger on the 5. Then move it
to the left 3 spaces. Your finger should end up on 2, so you
know that 5 − 3 = 2.

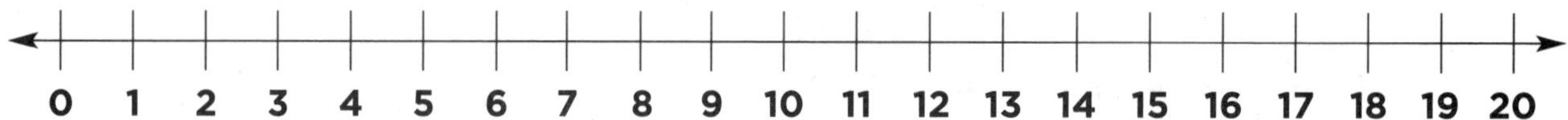

Place Value

Place value tells you the amount that each digit in a number stands for.

5

ones

5 ones = 5

15 = 1 + 5

tens   ones

1 ten = 10

5 ones = 5

---

215 = 2 + 1 + 5

hundreds   tens   ones

2 hundreds = 200

1 ten = 10

5 ones = 5

3,215 = 3 + 2 + 1 + 5

thousands   hundreds   tens   ones

3 thousands = 3,000

2 hundreds = 200

1 tens = 10

5 ones = 5

Multiplication Table

A multiplication table can help you practice the basic multiplication facts. To use a multiplication table, follow these steps:

- Find one of the numbers being multiplied in the first column.

- Find the other number being multiplied in the top row.

- Find the place where the row and column meet; that number is the product.

Example: $2 \times 5 = 10$

x	0	1	2	3	4	5	6	7	8	9
0	0	0	0	0	0	0	0	0	0	0
1	0	1	2	3	4	5	6	7	8	9
2	0	2	4	6	8	10	12	14	16	18
3	0	3	6	9	12	15	18	21	24	27
4	0	4	8	12	16	20	24	28	32	36
5	0	5	10	15	20	25	30	35	40	45
6	0	6	12	18	24	30	36	42	48	54
7	0	7	14	21	28	35	42	49	56	63
8	0	8	16	24	32	40	48	56	64	72
9	0	9	18	27	36	45	54	63	72	81

Rulers

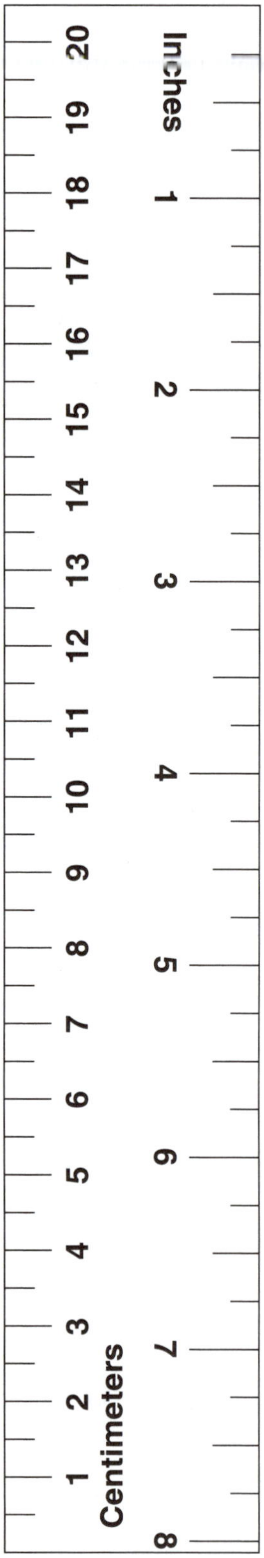

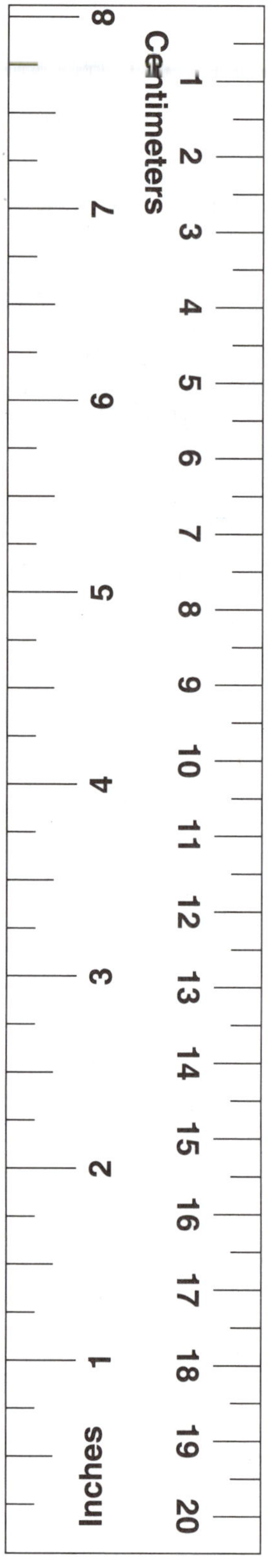

Geometic Figures and Polygons

Geometric Figures

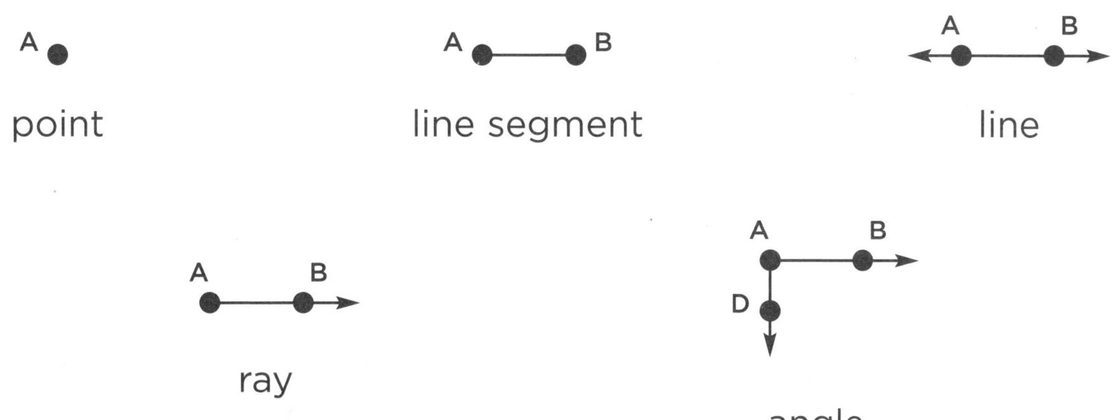

Quadrilaterals

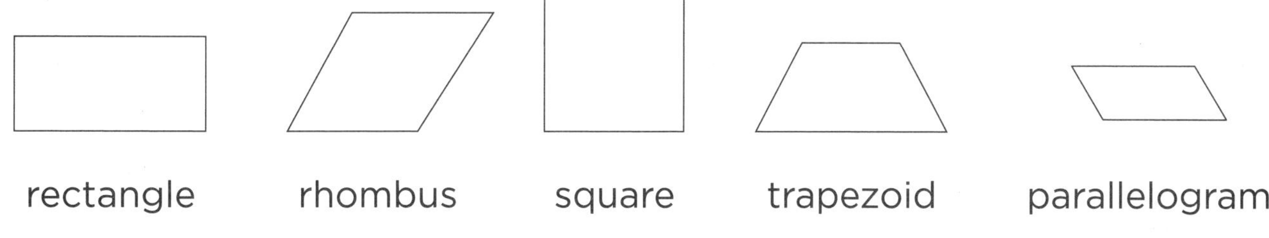

Other Polygons

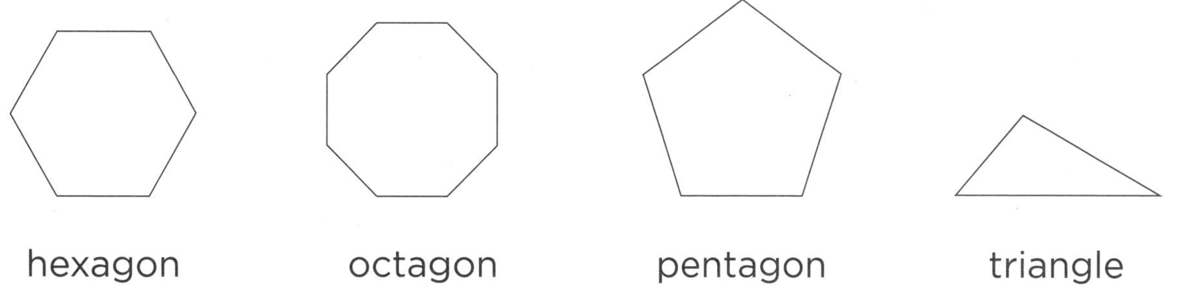

Fractions and Decimals

Fractions and decimals show parts of a whole or group.
The same value can be written as a fraction or decimal.
Some common amounts are shown below.

The rectangle is divided into four equal parts. Each part is $\frac{1}{4}$ or 0.25.

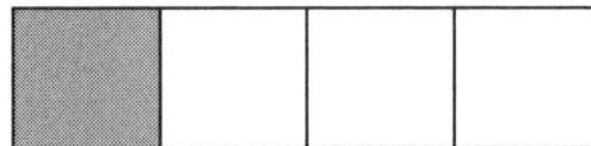

The circle is divided into two equal parts. Each part is $\frac{1}{2}$ or 0.5 or 0.50.

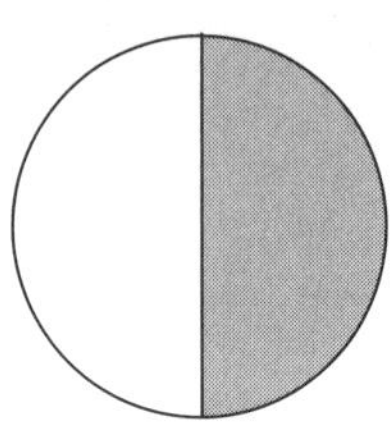

Equal in Value

In each chart, the values in a row are equal.

Fraction	Decimal
$\frac{1}{5}$	0.20 or 0.2
$\frac{2}{5}$	0.40 or 0.4
$\frac{3}{5}$	0.60 or 0.6
$\frac{4}{5}$	0.80 or 0.8
$\frac{5}{5}$	1.0

Fraction	Decimal
$\frac{1}{4}$	0.25
$\frac{2}{4}$	0.50 or 0.5
$\frac{3}{4}$	0.75
$\frac{4}{4}$	1.0